Comminution and Comminution Circuits Optimisation: 3rd Edition

Comminution and Comminution Circuits Optimisation: 3rd Edition

Guest Editors

Ngonidzashe Chimwani
Murray M. Bwalya

MDPI

Basel • Beijing • Wuhan • Barcelona • Belgrade • Novi Sad • Cluj • Manchester

Guest Editors
Ngonidzashe Chimwani
Mining, Minerals and
Geomatics Engineering
University of South Africa
Johannesburg
South Africa

Murray M. Bwalya
School of Chemical and
Metallurgical Engineering
University of the
Witwatersrand
Johannesburg
South Africa

Editorial Office
MDPI AG
Grosspeteranlage 5
4052 Basel, Switzerland

This is a reprint of the Special Issue, published open access by the journal *Minerals* (ISSN 2075-163X), freely accessible at: https://www.mdpi.com/journal/minerals/special_issues/GYKT7580ZZ.

For citation purposes, cite each article independently as indicated on the article page online and as indicated below:

Lastname, A.A.; Lastname, B.B. Article Title. *Journal Name* **Year**, *Volume Number*, Page Range.

ISBN 978-3-7258-4307-7 (Hbk)
ISBN 978-3-7258-4308-4 (PDF)
https://doi.org/10.3390/books978-3-7258-4308-4

Contents

About the Editors

Ngonidzashe Chimwani

Ngonidzashe Chimwani is a Senior Researcher in the Mining, Minerals, and Geomatics Engineering Department at UNISA, who holds a Ph.D. in Engineering (University of the Witwatersrand), an MBL (University of South Africa), and a B-Tech in Engineering (National University of Science and Technology). He is a C2 researcher, as rated by the National Research Foundation (NRF). Dr. Chimwani has received 11 awards throughout his career and has supervised several postgraduate students. His research focuses on optimising comminution circuits and extracting metals from low-grade ores and mineral processing waste dumps, using phytomining strategies for reuse and remediation purposes.

Murray M. Bwalya

Murray M. Bwalya is an Associate Professor in the Department of Chemical and Metallurgical Engineering, with a PhD from Witwatersrand university(RSA), MPhil Leeds University (UK) and BSc. from Cardiff University (UK), in the field of Mineral Processing. He has consulted widely as a Researcher mainly in South Africa and overseas. His area of expertise is comminution and has been involved in various aspects that include modelling, simulation and process optimization.

minerals

Editorial

Editorial for Special Issue "Comminution and Comminution Circuits Optimisation: 3rd Edition"

Ngonidzashe Chimwani

Department of Mining, Minerals and Geomatics Engineering, University of South Africa (UNISA), Florida Campus, Johannesburg 1710, South Africa; ngodzazw@gmail.com

Comminution is crucial to mining and mineral processing operations, as valuable minerals cannot be extracted if they are not liberated. This process spans from blasting to mineral processing, where the run-of-mine material is crushed and ground prior to separation. As articulated in this Special Issue, the efficiency of the separation stage, which determines the degree of mineral recovery, depends not only on the efficiency of the comminution stages but also on the understanding of the ore characteristics. However, achieving efficiency in comminution, which is widely known for a lion's share of the world's energy, has long been a mammoth task, given the non-revealing nature of comminution equipment, often referred to as "black boxes". The interactions among the operational parameters of the mills or crushers further complicate their understanding, making the use of techniques such as the discrete element method (DEM) helpful for optimising machines for size reduction. Considerable progress has been made in this area; nevertheless, the work is still far from exhaustive, especially considering global challenges such as environmental and energy crises, high-grade ore depletion, and unsustainable production. Thus, the research published in this Special Issue aims to address some of these challenges, providing important insights into blasting, ore pre-treatment and characterisation, optimal crushing and grinding, and the numerical assessment of comminution circuits.

In this vein, Contribution 1 used the Kuz–Ram model and Bond equation to assess the impact of surface blast design parameters—such as burden, spacing, stemming, and powder factor—on the fragmentation size and overall performance of a comminution circuit processing sedimentary copper-bearing ore, in terms of energy and throughput. The study resulted in a 20% increase in comminution circuit throughput, a 29% decrease in specific energy consumption, and a 12% total operating cost reduction, all of which contributed to improved overall fragmentation. Although higher powder factors were found to enhance processing, fragmentation that is too fine may lead to increased equipment wear and maintenance costs if an adequate balance is not maintained.

Focusing on the crushing stage in the comminution chain—in particular, the jaw crusher—Contribution 2 investigated the influence of mineralogical and physical properties on ore breakage behaviour. The authors explored the potential for the selective liberation of valuable minerals and determined that selective liberation was possible at relatively coarser particle sizes, which in turn improved ore beneficiation with lower energy consumption. They found that chalcopyrite and pentlandite exhibited better liberation at particle sizes below 400 μm, achieving over 50% liberation in the finest fraction, whilst magnetite was more enriched in particles smaller than 125 μm. The study emphasised the importance of considering mineralogical parameters when designing energy-efficient and selective comminution strategies.

A semi-autogenous (SAG) mill was assessed by Contribution 3, who aimed to characterise the stress and impact loading experienced by the grinding media in these mills

 https://doi.org/10.3390/min15050496

to help design stronger grinding media capable of withstanding stress and damage risk. The authors employed the DEM to evaluate the intensity of collisions using impact energy spectra (IES). The IES showed that higher ball loads resulted in more severe ball-on-ball impacts, increasing the risk of ball fracture. This study also showed that, while a higher mill speed is necessary to generate greater collision energy for ore breakage, it may also promote ball fracture. In SAG mills, bigger balls are preferred due to their high energy, which efficiently breaks particles; however, they are more vulnerable to fracture. Thus, the efficient operation of a SAG mill requires an optimum ball size to reduce ball damage, an optimum mill speed to balance ore breakage and media wear, and an optimum ball-to-ore ratio to balance ore cushioning and grinding efficiency.

Expanding on the milling stage of the mineral beneficiation chain, an integrated approach—combining a computer and the DEM coupled with the smoothed particle hydrodynamics (DEM-SPH)—was employed by Contribution 4 to calibrate the digital twin of a laboratory ball mill. They examined various milling scenarios, which included balls only, balls with ore, and balls with slurry, and simulated mill load behaviour. Their proposed method enabled the development of a numerical testing regime that can predict the grinding behaviour of the mill under different operational conditions without causing disruptive and costly changes in milling operations. The study combined experimental and simulation approaches, assessing the mill's behaviour under both dry and wet milling conditions, supported by image processing and the DualSPHysics framework. This research resulted in the creation of a calibrated digital twin, facilitating more efficient processing and energy utilisation during milling. Although the study was conducted using copper ore, the approach can be extended to other ores and milling environments. Designing optimised and energy-efficient milling operations is a critical step toward sustainable industrial practices, and this work provides a means to achieve that without the need for extensive physical testing.

Using vibrating disc mills, Contribution 5 investigated how the tap density of graphite is affected by grinding time. The study found that smaller particle sizes had higher tap densities than larger ones, while particles with a higher degree of sphericity resulted in lower tap densities. The authors attributed this behaviour to the impurities in the graphite, which affect packing and particle alignment during tapping. Consequently, graphite grinding using the vibratory disc mill can be optimised using exponential models, leading to energy and time savings.

Contribution 6 focused on optimising the rotational speed of the agitator in a vertical stirred mill, with the ultimate goal of enhancing the grinding efficiency. It is well established that increasing agitator speed enhances grinding efficiency, albeit with increased energy consumption. In this work, the authors successfully identified what they termed as the critical speed selection point, which balances grinding efficiency and energy consumption. This parameter is particularly significant, as even small efficiency gains can substantially reduce energy costs on an industrial scale. Notably, the evaluation index proposed in this study uses larger values to indicate greater grinding efficiency. From this perspective, an inverse proportional relationship was observed between rotational speed and grinding efficiency for a fixed milling time.

Several researchers have contributed to the optimisation of comminution circuits from the perspective of ore characteristics rather than comminution machinery. Contribution 7 established that the Bond Work Index can be applied to non-standard feed particle size distributions, with the help of a correction factor. This factor adjusts the work index to align with the standard feed conditions. For non-standard feed particles, the model was recommended for P100 = 75 μm tests.

Another method for measuring rock breakage energy and force is the Geopyörä Breakage Test, which uses two counter-rotating wheels to nip and crush rock samples between rollers with a tightly controlled gap. Contribution 8 employed this method and determined that it requires much smaller sample sizes and can accommodate broader, more detailed orebody characterisation.

An effective method for improving ore grindability is through pretreatment, as demonstrated by Contribution 9 who applied a hybrid of thermal and mechanical pretreatments before grinding. The Bond Work index tests revealed that microwave pretreatment alone significantly enhanced grindability compared to other individual treatment methods; however, prolonged microwave treatment time can cause iron ore particles to fuse, negatively affecting grindability. Mechanical pretreatment alone enhanced grindability by generating cracks, while furnace treatment was found to be the least effective. The combination of microwave and mechanical treatments yielded the best results in terms of grindability and energy savings. Nevertheless, a comprehensive energy assessment is necessary to account for the energy consumed during heating and compression.

In conclusion, the papers published in this Special Issue address interesting topics from various stages of comminution across the mineral beneficiation chain, ranging from blasting and crushing to milling. Ore pretreatment methods were also analysed, along with the grinding tests necessary for the optimisation of comminution circuits, providing valuable insights. The research presented in this Special Issue will be beneficial to both the mining and mineral processing industry and the academic community.

Acknowledgments: The Guest Editors thank the authors, reviewers, as well as the Editorial Board and staff from *Minerals* for their contributions to this Special Issue.

Conflicts of Interest: The author declares no conflicts of interest.

List of Contributions

1. Losaladjome Mboyo, H.; Huo, B.; Mulenga, F.K.; Mabe Fogang, P.; Kalenga Kaunde Kasongo, J. Assessing the Impact of Surface Blast Design Parameters on the Performance of a Comminution Circuit Processing a Copper-Bearing Ore. *Minerals* **2024**, *14*, 1226. https://doi.org/10.3390/min14121226.
2. Bravo, A.H.; Lieberwirth, H.; Popov, O. Evaluation of Ni-Cu Ore from Zapolyarnoe Based on Mineralogical and Physical Properties before and after Comminution. *Minerals* **2024**, *14*, 493. https://doi.org/10.3390/min14050493.
3. Bwalya, M.M.; Samukute, O.S.; Chimwani, N. SAG Mill Grinding Media Stress Evaluation—A DEM Approach. *Minerals* **2025**, *15*, 431. https://doi.org/10.3390/min15040431.
4. Doroszuk, B.; Bortnowski, P.; Ozdoba, M.; Krol, R. Calibrating the Digital Twin of a Laboratory Ball Mill for Copper Ore Milling: Integrating Computer Vision and Discrete Element Method and Smoothed Particle Hydrodynamics (DEM-SPH) Simulations. *Minerals* **2024**, *14*, 407. https://doi.org/10.3390/min14040407.
5. Güven, G.; Ulusoy, U.; Burat, F.; Mojtahedi, B.; Bayar, G. Time-Dependent Tap Density Modeling of Graphite Milled by Vibrating Disc Mill. *Minerals* **2025**, *15*, 403. https://doi.org/10.3390/min15040403.
6. Tang, B.; Cheng, B.; Song, X.; Ji, H.; Li, Y.;Wang, Z. Experimental Study on the Influence of Rotational Speed on Grinding Efficiency for the Vertical Stirred Mill. *Minerals* **2024**, *14*, 1208. https://doi.org/10.3390/min14121208.

7. Nikolić, V.; Pierres, J.M.; Calvo, M.S.; Menéndez-Aguado, J.M.; Trumić, M.; Trumić, M.S.; Milošević, V. Proposal of a Method for Calculating the Bond Work Index for Samples with Non-Standard Feed Particle Size Distribution. *Minerals* **2025**, *15*, 358. https://doi.org/10.3390/min15040358.
8. Bueno, M.; Almeida, T.; Lara, L.; Powell, M.; Delboni, H. Accuracy and Precision of the Geopyora Breakage Test. *Minerals* **2024**, *14*, 738. https://doi.org/10.3390/min14080738
9. Adewuyi, S.O.; Ahmed, H.A.M.; Anani, A.; Saeed, A.; Ahmed, H.M.; Alwafi, R.; Luxbacher, K. Enhancing Iron Ore Grindability through Hybrid Thermal-Mechanical Pretreatment. *Minerals* **2024**, *14*, 1027. https://doi.org/10.3390/min14101027.

minerals

MDPI

Article

Assessing the Impact of Surface Blast Design Parameters on the Performance of a Comminution Circuit Processing a Copper-Bearing Ore

Hervé Losaladjome Mboyo [1], Bingjie Huo [1,*], François K. Mulenga [2], Pieride Mabe Fogang [1] and Jimmy Kalenga Kaunde Kasongo [3,4]

1 School of Mining, Liaoning Technical University, Fuxin 123000, China; hervelosaladjome@gmail.com (H.L.M.); mabepieride2603@163.com (P.M.F.)
2 Department of Mining Engineering, University of South Africa, Florida Campus, Private Bag X6, Johannesburg 1710, South Africa; mulenfk@unisa.ac.za
3 Department of Mining Engineering, Polytechnic Faculty, University of Lubumbashi, Lubumbashi 1825, Democratic Republic of the Congo; jimmykalenga@gmail.com
4 Department of Geology and Mining Engineering, Polytechnic Faculty, Mapon University, Kindu 081, Democratic Republic of the Congo
* Correspondence: huobingjie@163.com; Tel.: +86-159-0498-6267

Abstract: Open-pit mining remains the dominant method for copper extraction in current operations, with blasting playing a pivotal role in the efficiency of downstream processes such as loading, hauling, crushing, and milling. This study assesses the impact of surface blast design parameters on the performance of a comminution circuit processing a copper-bearing ore. The analysis focuses on important design parameters such as burden, spacing, stemming, and powder factor, evaluating their influence on the fragment size distribution and downstream comminution circuit performance. Using the Kuz-Ram model, four novel blast designs are compared against a baseline to predict the size distribution of rock fragments (X80). Key performance indicators throughput and specific energy consumption are calculated to evaluate the comminution circuit performance. Results demonstrated that reducing the X80 from 500 mm to 120 mm led up to a 20% increase in throughput and a 29% reduction in total specific energy consumption. Furthermore, achieving finer particle sizes through more intensive blasting contributed to a reduction in total operating costs by up to 12%. These findings provide valuable insights for optimizing blast design to improve comminution circuit performance, contributing to sustainable mining practices by reducing energy consumption, operating costs, and the environmental footprint of mining operations.

Keywords: blast design parameters; copper bearing-ore; specific energy consumption; throughput; comminution; operating costs

1. Introduction

Rock fragmentation in mining involves breaking hard rocks into suitable sizes for easier downstream handling and processing. The most cost-effective and widely used method for ground fragmentation is currently drilling and blasting. This technique, which uses commercial explosives placed in blast holes to detonate a rock mass and break it into smaller pieces [1–3], has proven to be a reliable and efficient solution. Fragmentation is a critical outcome of rock blasting. It significantly impacts subsequent stages of the production cycle, such as loading, hauling, and downstream processes, including mineral processing. Good fragmentation enhances production efficiency and lowers the mining process's overall cost, reinforcing the current method's cost-effectiveness [4].

The results of a surface blast depend on several parameters, broadly categorized as controllable and uncontrollable [5,6]. The blasting engineer can adjust controllable parameters to influence the outcome of blasting operations. These parameters fall into three groups: geometric, explosives, and time-related. Geometric parameters encompass drill hole diameter, hole depth, charge length, spacing, burden, and stemming height. Explosives parameters include the type of explosive, explosive strength and energy, powder factor, and priming systems. Time parameters cover delay timing and initiation sequence [7]. The uncontrollable parameters consist of the geological and geotechnical properties of the rock mass. These parameters are inherent and cannot be adjusted to affect blasting outcomes. They include rock strength, rock-specific gravity, joint spacing and condition, presence and depth of water, and compressional stress wave velocity [5].

In hard rock mining, the ore is initially broken down into fragments of different sizes through blasting, also known as (blast) fragmentation. The fragments resulting from blasting are then further crushed and ground into smaller particles during crushing and grinding operations. Combined, crushing and grinding are referred to as comminution in mineral processing, with grinding being the most energy-intensive stage of the mineral processing [8] on most metal mine sites and a significant consumer in other commodities [9], accounting for 32% of the mine's energy consumption [10]. Comminution consumes a significant amount of energy, around 30%–60% of the total mine energy [11,12]; much of this energy is dissipated as heat into the rock. Comminution is one of the most expensive processes in mining and processing operations, both in terms of initial investment and ongoing operational expenses [13]. Effective comminution of ores is crucial for extracting valuable minerals and increasing operational efficiency [14]. Comminution circuit performance is usually assessed based on throughput, specific energy, and final grind size [15].

Since the 1990s, some studies have investigated the link between surface blast design parameters and downstream processes. The Mine-to-Mill (M2M) optimization approach, developed by the Julius Kruttschnitt Mineral Research Centre (JKMRC) in Australia, focuses on improving blasting activity to enhance productivity and energy efficiency in the production chain [12,16]. M2M optimization has been successfully implemented worldwide in various mining operations. Its application has increased throughput, reduced energy consumption, and cost [15,17–20]. The main goal of most studies was to improve mill performance. Hakami et al. [4] studied the effects of blast pattern design on autogenous and semi-autogenous mill throughput in an iron ore mine. By changing the blast pattern design from 5 m × 6 m to 4 m × 5 m, there was an increase in powder factor from 1650 g/m^3 to 2480 g/m^3, increasing mill throughput by 5%–30%. Dragano et al. [21] analyzed the connections between the geological environment, blasting parameters, and energy consumption during the primary crushing phase in a dolomitic limestone quarry. Their study aimed to understand the relationship between the energy provided for size reduction and the resistance to size reduction. Increasing the powder factor from 0.09 to 0.31 kg/t led to smaller particle sizes of 450–370 mm. Consequently, the energy consumption at the primary crusher decreases as the particle size (P80) decreases from 0.228 to 0.181 kWh/t. P80 decreases with the number of delays/t increases, and, finally, electricity cost decreases while the powder factor increases. Howe and Pan [22] used engineering research and industrial trial data to examine the connection between blasting fragmentation and comminution power consumption in a copper/gold open-cut porphyry deposit. They expanded the scope of their research to include the higher blasting powder factor range of 2–4 kg/m^3, which is achievable with Ultra-High-Intensity Blasting (UHIB) designs. Valery et al. [23] applied modeling and simulation of blasting and comminution to investigate the effects of blast design changes on downstream operations, aiming to increase throughput or improve product quality in iron and gold mines. Lam et al. [24] applied the M2M project to optimize blast design to improve the milling rate by altering feed size distribution at the Porgera gold mine. Ozdemir [25] examined the correlation between fragmentation size distribution values and jaw crusher throughput in a sandstone quarry.

Poor blast design often leads to suboptimal fragmentation, which increases the energy consumption of comminution circuits and limits throughput, ultimately raising operational costs. Although effective blast design can enhance comminution circuit performance, achieving the ideal fragment size involves trade-offs. A deeper understanding of the connection between specific blast design parameters and comminution performance metrics, such as throughput and specific energy consumption, is essential for achieving sustainable and cost-effective operations.

While the impact of blast design on comminution circuit performance is well-documented, most existing studies focus on porphyry copper deposits or mixed ore types, leaving a significant gap in understanding how blast parameters affect sedimentary or sandstone-hosted copper ores. Due to their distinct geological characteristics, these ores exhibit unique fragmentation and comminution behaviors, presenting different challenges for optimizing blast designs and enhancing downstream processes. This study addresses this gap by using predictive theoretical models to examine the effects of blast design parameters such as burden, spacing, stemming, and powder factor on comminution performance, specifically within sedimentary copper mining contexts.

This research contributes to academic knowledge and industry practice by addressing an underexplored area of blast-comminution optimization for sedimentary copper ores. Through theoretical modeling, it offers valuable initial insights into the distinct challenges and potential optimization strategies for these ore types. Academically, the findings advance the understanding of blast-comminution dynamics in these sedimentary-hosted copper mines. At the same time, from an industry perspective, they guide drill and blast engineers in developing tailored blast strategies to improve efficiency and cost-effectiveness in these specific geological settings, contributing to sustainable mining practices.

Optimizing surface blast design parameters, specifically burden, spacing, stemming, and powder factor, will reduce the X80 fragment size, thereby improving comminution circuit throughput and lowering specific energy consumption. Furthermore, excessively fine fragmentation may lead to higher wear rates, increased energy demand, and maintenance costs, offsetting potential benefits.

2. Materials and Methods

2.1. Case Study

The case study refers to an open-pit mine in the southern part of the Democratic Republic of the Congo (DRC). The mine primarily produces copper and silver as a by-product. This is an off-grid open-pit copper mine generating electricity on-site using diesel fuel. The mine's geological structure consists of folded and faulted sedimentary rocks trending northwest–southeast, hosting copper–cobalt mineralization within a syncline in the Central African Copperbelt region of the DRC. The mine deposit is found within a series of sedimentary rocks, mainly shales, sandstones, and conglomerates. Additionally, the ore deposit contains supergene (near-surface oxide and carbonate) and primary (sulfide) mineralization. The mineralogy of the oxide zone is mainly composed of malachite, azurite, clay alteration of feldspars, locally developed cuprite, native copper, and chalcocite. The primary sulfide mineral is chalcopyrite, with smaller amounts of bornite and chalcocite. The ore deposit has average grades of 3.6% copper (Cu) and 8.3 g/t silver (Ag).

The mine's conventional blast design, considered as a baseline, is presented in Table 1. A staggered pattern was used during the drilling and blasting operations, and the emulsion was the explosive type charged in the blast holes. The blast holes were drilled with a diameter of 125 mm and a depth of 11 m. Drilling cuttings were used as stemming material, with a sub-drilling depth of 1 m. The delay time between consecutive rows was 25 ms, and between holes in the same row was 17 ms. The bench height was 10 m. The blast holes were drilled vertically using the Drill rig roc L6 25 rotary machine, and the drilling operation was carefully performed with minimal deviation.

The baseline variables average values used in this study were derived from historical plant data, encompassing key variables such as fragment size distribution (X80), feed particle size (F80), product particle size (P80), throughput, specific energy consumption, and circulating load. These data, illustrated in Figure 1, offer foundational insights into comminution circuit performance, serving as a benchmark for evaluating the effects of different blast design parameters.

Table 1. Conventional blast design parameters.

Parameter	Unit	Value
Hole diameter	mm	125
Bench height	m	10
Burden	m	3–5
Spacing	m	3–5.5
Sub-drilling	m	1
Hole depth	m	11
Stemming height	m	Variable
Specific gravity	g/cc	2.6
Explosive	-	Emulsion P100
Explosive density	g/cc	1.15
Velocity of Detonation (VoD)	m/s	5000
Relative Weight Strength (RWS) (ANFO = 100)	%	85
Powder factor	kg/m^3	0.12–0.55

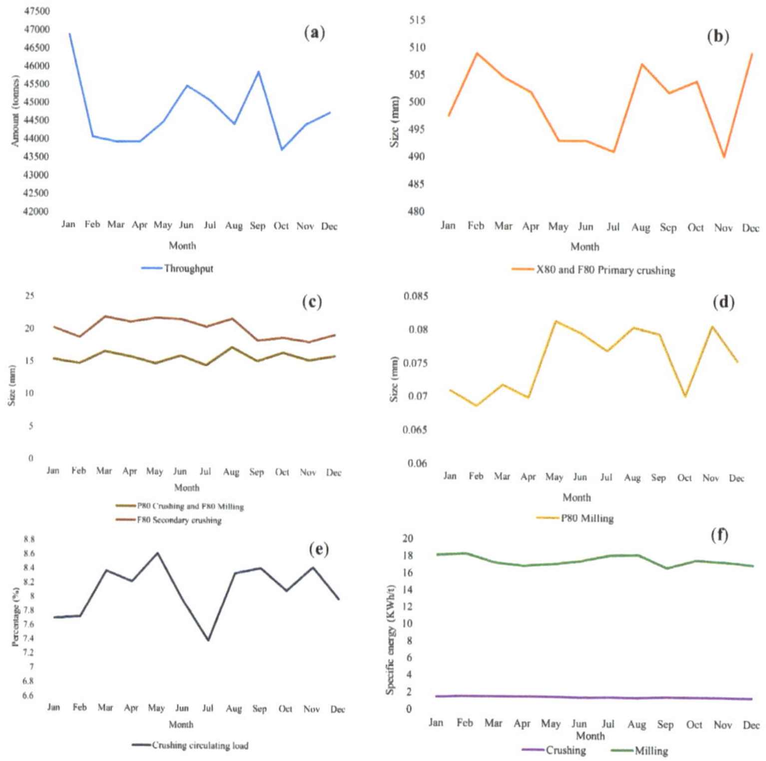

Figure 1. Monthly averages of main historical variables over the two-year production period from 2021 to 2022: (**a**) Throughput, (**b**) X80 and F80 Primary crushing, (**c**) P80 Crushing and F80 Milling, and F80 Secondary crushing, (**d**) P80 Milling, (**e**) Crushing circulating load and (**f**) Crushing and milling specific energy consumption.

2.2. Research Methodology

This study explores the impact of surface blast design parameters on the performance of a comminution circuit processing a copper-bearing ore. The methodology is grounded in the Mine-to-Mill (M2M) approach, which seeks to optimize upstream blasting operations to enhance the efficiency and cost-effectiveness of downstream processes, particularly crushing and milling. The M2M approach uses intense blasting, achieved by increasing the powder factor and simultaneously adjusting multiple parameters to generate finer fragmentation. Adjustments to the powder factor were made to capture a range of fragmentation outcomes representative of typical operational scenarios in hard rock mining. The selected variations in the powder factor served two key objectives. First, they aimed to assess the impact on fragment size distribution, as changes in the powder factor directly influence the energy applied per unit volume, which determines the X80 size and affects downstream comminution performance. Second, they were intended to explore energy-performance trade-offs. Higher powder factors, which produce finer fragmentation, can enhance throughput and reduce energy consumption, while lower powder factors, though more cost-effective in terms of explosives, may negatively impact circuit performance. This study investigates how variations in key blast parameters affect comminution circuit performance. To quantify these effects, theoretical models, such as the Kuz-Ram model for predicting fragment size distribution (X80) and the Bond equation for energy consumption, were used to establish the relationships between blast design parameters and comminution circuit performance measured by two metrics, throughput, and specific energy consumption.

Four alternative blast designs are proposed, with changes made to key parameters such as burden, spacing, stemming height, and powder factor. Specifically, burden was reduced by 13%–38%, spacing by 11%–33%, and stemming height by 25%–75%, while the powder factor increased by 47%–242%, as shown in Table 2. All other parameters of the conventional blast design remained unchanged. Figure 2 shows the blast pattern for the baseline scenario, which served as the benchmark for comparison. The proposed blast designs were evaluated against the conventional design, using blasting outcomes to assess their impact on comminution circuit performance.

Table 2. Blast design parameters changes.

Parameter	Baseline	Design 1	Design 2	Design 3	Design 4
Burden (m)	4	4	3.5	3	2.5
Spacing (m)	4.25	4.50	4	3.5	3
Stemming height (m)	Variable	4	3	2	1
Powder factor (kg/m^3)	0.12–0.55	0.55	0.81	1.21	1.88

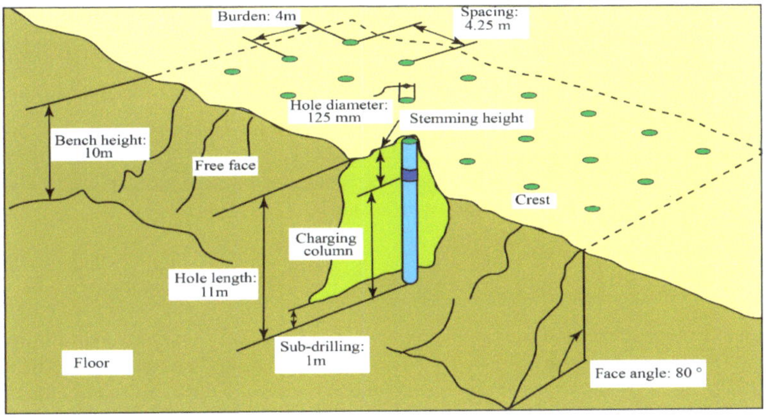

Figure 2. Baseline blast pattern.

Finally, a comprehensive cost analysis was performed to quantify the impact of the proposed blast designs on drilling, blasting, and comminution operating costs. The baseline operating cost estimates were compared with those of the alternative blast designs to highlight variations in cost outcomes resulting from the adjusted parameters.

2.2.1. Kuz–Ram Model Analysis

Kuz-Ram model is widely used to determine the size of fragments before blasting. This model considers the particle size distribution of the resulting fragments, the 50% of the fragment size distribution (X50), as the average size of the generated blocks, and the uniformity index (n). The Kuz–Ram model includes three equations: the modified Kuznetsov equation (Equation (1)), first developed by Kuznetsov [26] and later modified by Cunningham [27]; the Rosin–Rammler equation (Equation (2)) [28,29]; and the Cunningham uniformity index (n) equation (Equation (3)) [30]. It is important to note that the K-factor is also referred to as the powder factor, which is used to assess the rock fragmentation power for the desired granulometry as follows:

$$X_{50} = AK^{-0.8}Q^{\frac{1}{6}}\left(\frac{115}{RWS}\right)^{\frac{19}{30}} \quad (1)$$

where X_{50} is the average size at which 50% of the particles pass the sieve (cm); A is the characteristic factor of the rock; K is the powder factor (kg/m^3); Q is the mass of explosive being used per hole (kg); and RWS is the relative power by weight of the explosive relative to the energy of the ANFO (115 is the RWS of the trinitrotoluene (TNT)) (%).

The Rosin–Rammler formula predicts the fragment size distribution. It is generally recognized as providing a reasonable description of fragmentation in blasted rock based on the percentage of retained fragments for a sieve opening size [28]:

$$R_m = 1 - e^{-\left(\frac{X}{X_c}\right)} \quad (2)$$

where R_m is the cumulative frequency passing through the opening of the sieve of size (%), X is the size of the sieve opening (cm), and X_C is the characteristic size (cm). If the characteristic size X_C, through which 63.2% of the particles pass, and the uniformity index are known, X_C can be obtained using the following equation [31]:

$$X_c = \frac{X_{50}}{\sqrt[n]{0.693}} \quad (3)$$

Cunningham [29] developed an equation to estimate the uniformity coefficient "n" of the Rosin–Rammler distribution curve from blast design parameters:

$$n = \left(2.2 - 14\frac{B}{d}\right)\left(\frac{1}{2} + \frac{S}{2B}\right)^{0.5}\left(1 - \frac{W}{B}\right)\left(\frac{L}{H}\right) \quad (4)$$

where B is the burden (m), S is the spacing (m), d is the blast hole diameter (mm), W is the standard deviation of drilling accuracy (m), L is the total charge length (m), and H is the bench height (m). Where there are two different explosives in the blast hole (bottom charge and column charge) [31], Equation (4) is modified to:

$$n = \left(2.2 - 14\frac{B}{d}\right)\left(1 - \frac{W}{B}\right)\left(\frac{1}{2} + \frac{S}{2B}\right)^{0.5}\left(\text{abs}\left(\frac{BCL - CCL}{L}\right) + 0.1\right)^{0.1}\left(\frac{L}{H}\right) \quad (5)$$

where BCL is the bottom charge length (m), and CCL is the column charge length (m). This equation should be multiplied by 1.1 when using a staggered pattern. The value of n is crucial in defining the shape of the Rosin–Rammler curve and typically ranges from (0.7 to 2). Higher values indicate uniform sizing, while lower values indicate a broad range of sizes, including oversize and fines [30].

Cunningham [30] modified the model twenty years later, mainly due to the introduction of electronic delay detonators. This resulted in what is now known in the literature as the modified Kuz-Ram model. Cunningham's adjustments consider the effects of inter-hole delay, timing scatter, and correction factors for the rock factor and uniformity index. These changes led to the modification of Equations (1) and (5) as follows [30]:

$$X_{50} = AA_T K^{-0.8} Q^{1/6} \left(\frac{115}{RWS}\right)^{19/20} C(A) \quad (6)$$

$$n = n_s \sqrt{\left(2 - \frac{30B}{d}\right)} \sqrt{\left(\frac{1+\frac{S}{B}}{2}\right)} \left(1 - \frac{W}{B}\right) \left(\frac{L}{H}\right)^{0.3} C(n) \quad (7)$$

where A_T is a timing factor for the effect of inter-hole delay, C(A) represents a correction factor for the rock factor, n_s is the uniformity factor for the impact of timing scatter, and C(n) is a correction factor for the uniformity index. Therefore, the modified Kuz–Ram model consists of Equations (2), (6), and (7). This model was used in this study to predict the particle sizes of the blast designs.

2.2.2. Comminution Circuit Performance

The comminution circuit of the study case comprised two stages: crushing (primary and secondary crushers) and grinding (two ball mills). The cyclone overflow is sent to the flotation plant to extract the copper. This comminution circuit has a nominal throughput of 2400 tons/day. It comprised a jaw crusher, a cone crusher, and two ball mills, producing a final product particle size (P80) of 75 microns (μm). Figure 3 below shows the comminution circuit flowsheet.

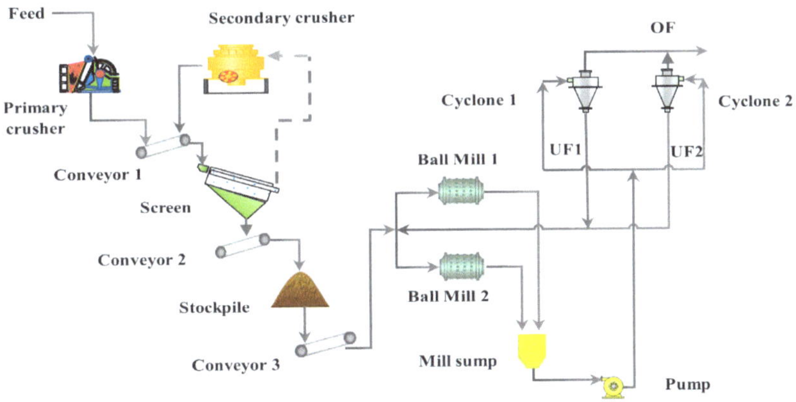

Figure 3. Comminution circuit flowsheet. UF is underflow and OV is overflow.

In this study, Equation (8) from Figueiredo et al. [32], adapted for the present case study, was used to estimate the productivity of primary crushing (throughput) based on the blast fragmentation parameter X80:

$$P_{bpX80} = P_{bp} e^{-0.002X80} \quad (8)$$

where P_{bpX80} is the primary crushing productivity (t/h), and P_{bp} is the primary crushing nominal productivity (t/h).

The Bond equation (Equation (9)) [33,34] was used to calculate the specific energy of the comminution circuit, which is required to grind the ore from a known feed size (F80) to a desired product size (P80):

$$E = 10 \times W_i \left(\frac{1}{\sqrt{P_{80}}} - \frac{1}{\sqrt{F_{80}}}\right) \quad (9)$$

where E is the specific energy (kWh/t), W_i is the Work index and is a measure of the ore's resistance to crushing and grinding (kWh/t) [33], P_{80} is the product particle size (μm), and F_{80} is the feed particle size (μm). One reason for using Bond's third theory is that the work index W_i has been measured and reported for many rocks.

Based on historical plant data (Figure 1), the following outlines the particle size variations in the feed and product streams at each stage of the comminution circuit:

- Crushing circuit: For each of the four designs, the X80 values, obtained via the Kuz-Ram model analysis, represented the F80 for the primary crusher. In the secondary crusher, the F80 represented the circulating load size, the portion of material with particle sizes larger than the screen aperture, which failed to pass through the screen. The combined fraction of ore from both the primary and secondary crushers, with particle sizes below the screen aperture, established the P80 for the overall crushing circuit;
- Milling circuit: The P80 from the crushing circuit served as the F80 for the milling circuit in both ball mills. The resultant particle size from the milling process defined the P80 for the milling circuit.

2.2.3. Operating Costs Estimation

The impact of blast design parameters extends beyond blasting itself and significantly influences not only comminution performance but also operating costs. The methodology for estimating operating expenses (OPEX) was based on production and operational data, direct operating expenditures, and associated annual costs for various components involved in operations such as drilling, blasting, and comminution.

The primary components of operating costs comprised equipment depreciation, labor, fuel, maintenance, consumables, and contingency. The total cost for each operation was obtained by summing the expenses across these components, and the cost per ton was calculated using annual production data. A 10% contingency factor was applied to the total operating cost to account for unforeseen expenses.

- Drilling

Annually, 5,274,076 tons of material were drilled, requiring 14,038 blast holes. The following equations were used to estimate the drilling operation's operating cost:

$$C_t = C_{Ed} + C_{Lab} + C_{MR} + C_{Fl} + C_{DCons} + C_{Oth} \tag{10}$$

$$OPEX = \frac{C_{tc}}{MD} \tag{11}$$

where C_t is the total cost (USD/year), C_{Ed} is the equipment depreciation cost (USD/year), C_{Lab} is the labor cost (salaries and wages for drill operators and technicians: USD/year), C_{MR} is the maintenance and repairs cost (USD/year), C_{Fl} is the fuel cost (USD/year), C_{DCons} is the drilling consumables and repairs cost (USD/year), C_{Oth} is the other costs (USD/year), $OPEX$ is the drilling operating cost (USD/ton), C_{tc} is the total cost with a 10% contingency (USD/year), and MD is the material drilled (tons/year).

Moreover, when calculating the $OPEX$ for the alternative designs, only the MD was changed, with all other cost components assumed to remain constant.

- Blasting

Annually, 6,316,946.65 tons of material were blasted, including 232,300.92 tons of ore and 6,084,645.73 tons of waste. The total charge mass reached 1009.32 tons per year. The following equations were used to estimate the blasting costs:

$$C_t = C_{Lab} + C_{MS} + C_{Fl} + C_{PR} + C_{Oth} \tag{12}$$

$$OPEX = \frac{C_{tc}}{MB} \tag{13}$$

where C_{Lab} is the labor cost (salaries and wages for blast operators and technicians: USD/year), C_{MS} is the materials and supplies cost (explosives and accessories: USD/year), C_{PR} is the permits and regulatory cost (blasting licenses, insurances and consumables cleaning: USD/year), *OPEX* is the blasting operating cost (USD/ton), and *MB* is the material blasted (tons/year).

Furthermore, when calculating the *OPEX* for the alternative designs, only the *MB* and the cost of materials and supplies (explosives and accessories) were modified, while all other cost components were assumed to remain unchanged.

- Comminution

The comminution circuit achieved an average annual throughput of 544,438 tons. Given the similarity in cost components between crushing and grinding, the operating costs for these processes were estimated using the following equations:

$$C_t = C_{Ed} + C_{Lab} + C_{MR} + C_{Ec} + C_{CCons} + C_{Oth} \tag{14}$$

$$OPEX = \frac{C_{tc}}{TPH} \tag{15}$$

where C_{Lab} is the labor cost (salaries and wages for comminution operators and technicians: USD/year), C_{Ec} is the energy consumption cost (crushing or milling: USD/year), C_{CCons} is the comminution (crushing or milling) consumables cost (USD/year), *OPEX* is the operating cost (crushing or milling: USD/ton), and *TPH* is the throughput (tons/year).

In calculating the OPEX for the alternative designs, only throughput and energy consumption were adjusted, while all other cost components, such as maintenance and repairs and consumables, remained constant. These components were not estimated due to the complexity of their input data and parameters. Although particle size and energy consumption were reduced, their associated costs may remain unchanged or even increase with higher throughput.

3. Results and Discussion

3.1. Blast Fragmentation

A Run-of-Mine (ROM) fragmentation distribution analysis was conducted using the Kuz-Ram model, applying the newly established blast design parameters. This model, defined by Equations (2)–(4) and (6), enabled the prediction of particle size distribution post-blasting. The analysis results are depicted in Figures 4 and 5, with a summary in Table 3. The fragmentation data have been graphically represented using a logarithmic scale to understand the blasting outcomes better. This approach highlights the distribution trends more effectively, particularly in the finer and coarser fractions. It allows for a more nuanced interpretation of the blasting performance and its subsequent impact on downstream processes.

The particle size distribution curves and the data presented in Table 3 reveal significant findings when applying the Kuz-Ram model to the various blast designs. Specifically, designs 1–4 yielded an X80, the size at which 80% of the material passes, within a range of approximately 120 to 320 mm with the new modified parameters. This is notably smaller than the conventional blast design (baseline), where the X80 value could reach up to 500 mm, the maximum fragment size that the primary crusher can accommodate. These results indicate that the proposed modifications, which implement a smaller blast pattern and increase the powder factor, have successfully improved fragmentation. The reduced X80 values for designs 1–4 demonstrate more efficient rock breakage, leading to smaller and more manageable fragment sizes than those produced by the baseline conventional blast. The findings are consistent with some studies highlighting the potential for optimizing ROM fragmentation through strategic changes in blast design [15,35,36]. By refining the blast parameters, it is possible to enhance downstream processing efficiency and potentially improve overall operational productivity.

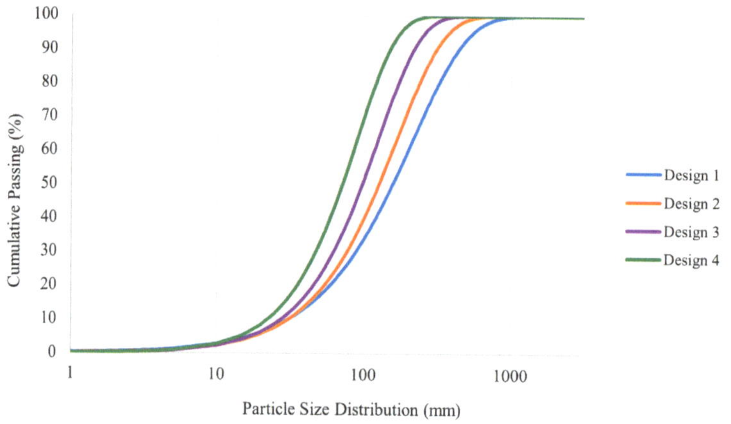

Figure 4. Kuz-Ram particle size distribution.

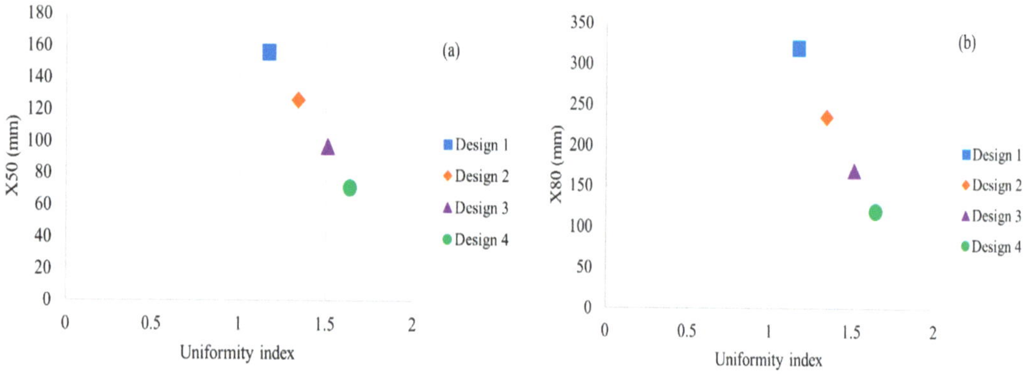

Figure 5. Relationship between uniformity index and particle size distribution: (**a**) n relative to X50, (**b**) n relative to X80.

Table 3. Parameters obtained using the Kuz–Ram model.

Parameter	Design 1	Design 2	Design 3	Design 4
n	1.17	1.34	1.51	1.64
X_c (mm)	213	166	124	89
X50 (mm)	156	126	97	71
X80 (mm)	320	236	170	120
X100 (mm)	888	572	375	247

The particle size distribution analysis revealed a consistent increase in the uniformity index across the four new blast designs, with values ranging from 1.17 to 1.64. This increase is inversely related to the X50 and X80 values, as Figure 5 illustrates. The rising uniformity index suggests that the size distribution of the resulting fragments becomes more consistent, with a narrower range of fragment sizes produced during the breakage process. In other words, the fragments are more homogeneously sized, with less variation between the smallest and largest pieces [37]. A higher uniformity index indicates a more favorable fragmentation pattern, which can significantly benefit subsequent processing stages. With more uniformly sized fragments, the efficiency of crushing and grinding operations can be enhanced, potentially leading to improved overall processing performance and reduced energy consumption.

The results of blast design optimization have significant implications for enhancing efficiency and reducing costs in real-world mining operations. By demonstrating how different blast designs affect fragmentation and energy consumption, mining planning engineers can strategically implement these insights to streamline processes such as loading, hauling, and crushing. For instance, optimized fragmentation can lead to faster loading times and reduced wear on crushing equipment, thereby decreasing operational downtime and maintenance costs. Furthermore, by adopting designs that lower specific energy consumption, mines can achieve substantial cost savings on energy bills while simultaneously minimizing their environmental footprint.

3.2. Comminution Circuit

3.2.1. Throughput

Table 4 and Figure 6 present the productivity estimates for the primary crushing, calculated based on the X80 values obtained from each alternative blast design using Equation (8). The X80 value, representing the screen size through which 80% of the ROM material passes, serves as a critical indicator of the effectiveness of the blast design in optimizing downstream processes. Based on historical data, the baseline productivity is 544,438 dry tons annually, translating to 45,368 tons per month, 1578 tons per day, and 66 tons per hour, with 345 operating days (8280 h) per year. The baseline case used historical production data as a benchmark to evaluate the performance indicators of four alternative designs.

Table 4. Estimation of primary crushing productivity.

Parameters	Design 1	Design 2	Design 3	Design 4
P_{bp} (t/h)	100	100	100	100
X80 (mm)	320	236	170	120
P_{bpX80} (t/year)	438,840	513,360	587,880	654,120
P_{bpX80} (t/day)	1272	1488	1704	1896
P_{bpX80} (t/h)	53	62	71	79
Variation (%)	−20	−6	+8	+20

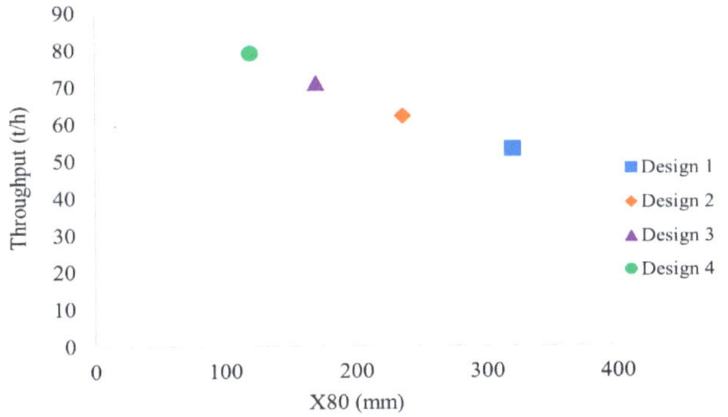

Figure 6. Primary crushing productivity relative to X80.

As anticipated, the productivity of primary crushing improved with the reduction in blasting fragment size, as indicated by the X80 values for the four alternative blast designs. However, when comparing the estimated productivity of the new blast designs to the baseline, designs 1 and 2 exhibited lower productivity than the baseline. This outcome may be attributed to the blast pattern parameter values used in the baseline, particularly the spacing, which was slightly smaller in Design 1 and larger in Design 2 (Table 2).

A closer examination of the productivity estimates, as shown in Figure 7, highlights the direct correlation between the increase in powder factor and the rise in throughput.

This suggests that the actual X80 value for the baseline case likely falls between the X80 values of designs 2 and 3. With the adjustments in blasting practices, the productivity of primary crushing could potentially be enhanced by up to 20% in this case study (Table 4).

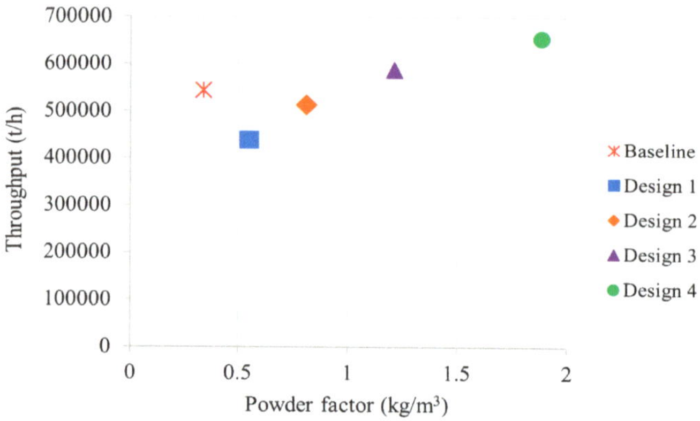

Figure 7. Primary crushing productivity relative to powder factor.

This finding is particularly relevant for hard rock mines, such as the one under study, and aligns with the results reported by other researchers who have explored the effect of surface blast design parameters on primary crushing productivity [4,15,35,36]. The potential for a 20% increase in productivity highlights the significant benefits of optimizing blast design to improve overall processing efficiency.

3.2.2. Energy Consumption

Tables 5 and 6 present the calculated specific energy consumption for the primary and secondary crushing circuits. Equation (9) estimated the specific energy consumption for the four alternative blast designs, whereas the baseline specific energy consumption for the entire crushing circuit, derived from historical data, is 1.55 kWh/t.

Table 5. Primary crushing specific energy consumption.

Parameter	Design 1	Design 2	Design 3	Design 4
F80 (μm)	320,000	236,000	170,000	120,000
P80 (μm)	16,000	16,000	16,000	16,000
W_i (kWh/t)	12	12	12	12
E (kWh/t)	0.74	0.70	0.66	0.60

Table 6. Secondary crushing specific energy consumption.

Parameter	Design 1	Design 2	Design 3	Design 4
F80 (μm)	20,000	20,000	20,000	20,000
P80 (μm)	16,000	16,000	16,000	16,000
W_i (kWh/t)	12	12	12	12
E (kWh/t)	0.1002	0.1002	0.1002	0.1002
Throughput (t/h)	4	5	6	6
Throughput (t/year)	35,107	41,069	47,030	52,330

The specific energy consumption for primary crushing decreased from 0.74 to 0.60 kWh/t, driven by variations in feed particle size (F80) across the four alternative blast designs (Table 5). These F80 values correspond to the X80 derived from the Kuz-Ram model analysis (Table 3), while the product size (P80) and the work index remained constant based on historical plant data. In contrast, the specific energy consumption for secondary crushing remained constant at 0.1002 kWh/t throughout the four alternative blast designs, as F80 and P80 values remained unchanged (Table 6). The F80 in secondary crushing represents only the oversized ore (circulating load) that did not pass through the screen after primary crushing, which is calibrated to pass ore up to 16,000 μm. Based on historical plant data, with a circulating load of approximately 8%, the productivity across the new four-blast designs fluctuated around 5 t/h. When evaluating the total specific energy consumption for both primary and secondary crushing, a significant reduction was observed, decreasing from 0.8402 to 0.7002 kWh/t (Table 7). This represents a 46%–55% decrease for the four alternative blast designs compared to the baseline value of 1.55 kWh/t. These results demonstrated that increasing the powder factor in blasting operations resulted in smaller particle sizes and reduced energy consumption within the crushing circuit.

Table 7. Total crushing specific energy consumption.

Parameter	Baseline	Design 1	Design 2	Design 3	Design 4
X80 (mm)	500	320	236	170	120
Specific energy (kWh/t)	1.55	0.8402	0.8002	0.7602	0.7002
Decrease (%)	-	46	48	51	55

Table 8 and Figure 8 summarize the relationship between X80, crushing productivity, and specific energy consumption. Reducing X80 is inversely related to productivity but directly tied to lowering energy consumption.

Table 8. Relationship between fragmentation particle size and crushing performance.

Parameter	Baseline	Design 1	Design 2	Design 3	Design 4
X80 (mm)	500	320	236	170	120
Throughput (t/h)	66	53	62	71	79
Specific energy (kWh/t)	1.55	0.8402	0.8002	0.7602	0.7002

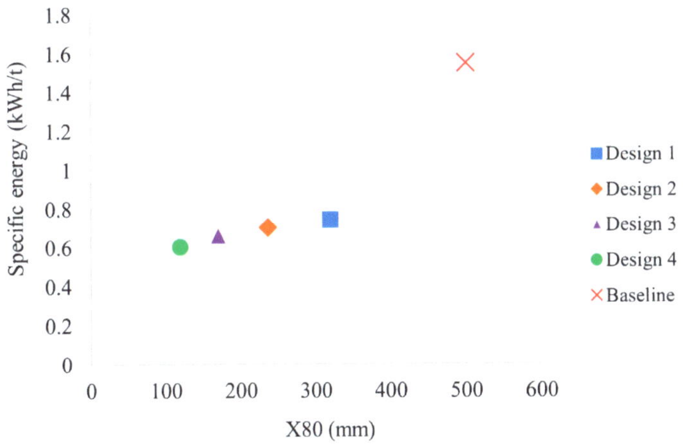

Figure 8. Crushing specific energy related to X80.

Table 9 provides the specific energy consumption calculated for the milling circuit, consisting of two ball mills operating in dry mode. As with the crushing circuit, Equation (9) estimated the specific energy consumption for each of the four alternative blast designs. However, the baseline specific energy consumption for the milling circuit, derived from historical data, is 17.62 kWh/t.

Table 9. Milling specific energy consumption.

Parameter	Design 1	Design 2	Design 3	Design 4
F80 (μm)	16,000	16,000	16,000	16,000
P80 (μm)	75	75	75	75
W_i (kWh/t)	12	12	12	12
E (kWh/t)	12.91	12.91	12.91	12.91
Throughput (t/h)	53	62	71	79
Throughput (t/year)	438,840	513,360	587,880	654,120

The total milling-specific energy consumption decreased significantly from 17.62 to 12.91 kWh/t, representing a 27% reduction across the four alternative blast designs compared to the baseline case. This reduction is particularly noteworthy given that grinding is the most energy-intensive stage in mineral processing, especially in hard rock mining operations [8]. Additionally, the F80 and P80 sizes remained constant across the four alternative blast designs, as these values are sourced from historical plant data and incorporated into the calculations for milling-specific energy consumption using Equation (9). The F80 value represents only the ore output from the crushing circuit (P80), calibrated to a maximum particle size of 16,000 μm, and serves as the feed for the milling circuit, based on historical plant data used for this study. The findings could suggest that the proposed blast modifications can significantly enhance operational efficiency by lowering energy costs without compromising productivity.

3.2.3. Comminution Circuit Performance

Table 10 summarizes the study's findings on the relationship between surface blast design parameters and the performance of a comminution circuit that processes copper-bearing ore.

Table 10. Comminution circuit performance summary.

Parameter	Baseline	Design 1	Design 2	Design 3	Design 4
X80 (mm)	500	320	236	170	120
Throughput (t/h)	66	53	62	71	79
Variation (%)	-	−20	−6	+8	+20
Specific energy (kWh/t)	19.17	13.75	13.71	13.67	13.61
Decrease (%)	-	28	28	29	29

Figure 9 presents the relationship between X80, throughput, and specific energy consumption of the comminution circuit. It is evident that with the changes made to the different blast designs, the X80 decreases, throughput increases and specific energy decreases from baseline to Design 4.

The performance of a comminution circuit is typically assessed using two key metrics: throughput (t/h) and specific energy consumption (kWh/t). This study strongly correlated blast design parameters and comminution circuit performance. Notably, a reduction in X80, the particle size at which 80% of the material passes through the primary crusher, led to a throughput increase of up to 20% (Table 4) and a significant reduction in total comminution circuit specific energy consumption by up to 29% (Table 10). These results are consistent with findings from other studies, reinforcing the critical role of blast design optimization in enhancing the overall performance of the mining value chain [36,38].

Even though the theoretical estimates used in the analysis of comminution circuit performance provided valuable insights, they should be interpreted cautiously. These correlations are based on established theories, principles, or assumptions that may not apply universally across all mining scenarios. Thus, while they offer useful approximations, they should not be viewed as definitive predictors. To enhance their validity, it is essential to supplement these estimates with data from controlled experiments or pilot studies tailored to the specific mining context. Additionally, conducting sensitivity analyses can help assess the impact of parameter variations, providing a more nuanced understanding of their applicability. Ultimately, conclusions drawn from these estimates should be contextualized within the limitations of the underlying data and assumptions.

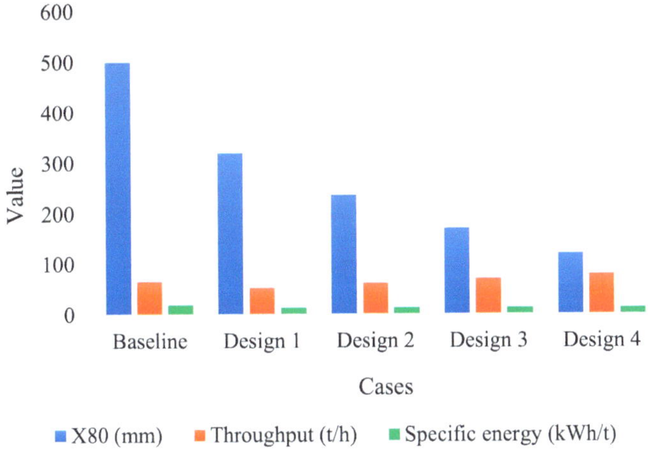

Figure 9. Relationship between X80, throughput, and specific energy.

3.3. *Operating Costs Estimation: Baseline Case*

3.3.1. Drilling

Table 11 presents the estimated drilling operating costs calculated from Equations (10) and (11). Additionally, Figure 10 provides a visual representation of the cost distribution among the various components contributing to the total operating cost.

Table 11. Drilling operating cost estimation.

Components	Cost (USD/Year)
Equipment depreciation	83,160.00
Labor	127,650.00
Maintenance and Repairs	567,210.90
Fuel	532,001.39
Consumables	130,036.90
Other	209,481.89
C_t	1,649,541.07
Contingency (10% C_t)	164,954.11
C_{tc}	1,814,495.18
OPEX (USD/t)	0.34

The estimated operating cost of drilling is primarily driven by maintenance and repairs, accounting for about 31% of the total. Fuel is another significant contributor at approximately 29%. Together, these two components represent the bulk of the expenses. In contrast, consumables, labor, and equipment depreciation are the least significant cost drivers, comprising around 7%, 7%, and 5%, respectively. Their combined impact remains minimal compared to the substantial contributions of maintenance and fuel.

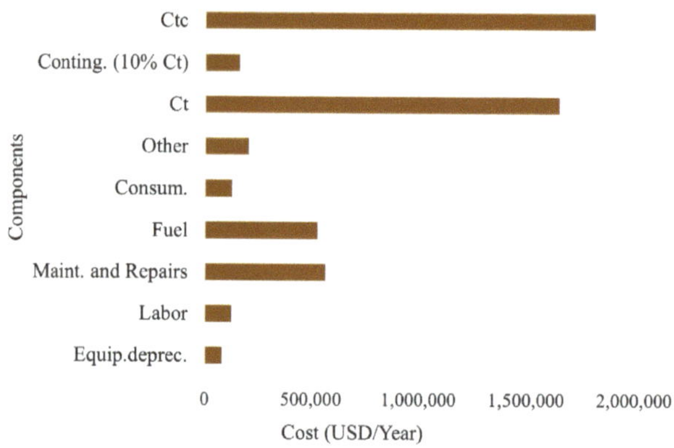

Figure 10. Drilling operating cost components distribution.

3.3.2. Blasting

Table 12 summarizes the estimated drilling operating costs, calculated using Equations (12) and (13). In addition, Figure 11 provides a breakdown of the cost distribution across the various components contributing to the total operating cost.

Table 12. Blasting operating cost estimation.

Components	Cost (USD/Year)
Labor	55,200.00
Explosives and Accessories	4,768,272.93
Fuel	399,001.04
Permits and Regulatory	403,525.46
Other	209,481.89
C_t	5,835,481.32
Contingency (10% C_t)	583,548.13
C_{tc}	6,419,029.45
OPEX (USD/t)	1.02

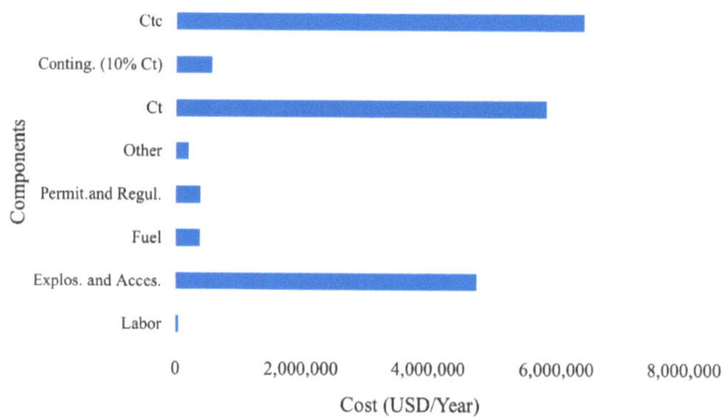

Figure 11. Blasting operating cost components distribution.

Explosives and accessories dominate the estimated blasting operating costs, accounting for approximately 74% of the total cost, highlighting their role as the primary cost driver. Fuel, along with permits and regulatory fees, each contributes about 6%, while labor accounts for only 1%.

3.3.3. Comminution

Equations (14) and (15) were used to estimate the comminution operating costs, which are presented in Tables 13 and 14. Figures 12 and 13 visually depict the cost distribution across the various components involved in the operating cost calculations for crushing and milling.

Table 13. Crushing operating cost estimation.

Components	Cost (USD/Year)
Equipment depreciation	28,948.53
Labor	853,329.01
Maintenance and Repairs	462,755.57
Energy consumption	1,167,495.57
Consumables	797,140.91
Other	184,812.56
C_t	3,494,482.15
Contingency (10% C_t)	349,448.22
C_{tc}	3,843,930.37
OPEX (USD/t)	7.06

Table 14. Milling operating cost estimation.

Components	Cost (USD/Year)
Equipment depreciation	62,088.75
Labor	2,346,654.77
Maintenance and Repairs	925,511.14
Energy consumption	3,891,651.91
Consumables	3,065,926.58
Other	736,975.18
C_t	11,028,808.33
Contingency (10% C_t)	1,102,880.83
C_{tc}	12,131,689.16
OPEX (USD/t)	22.28

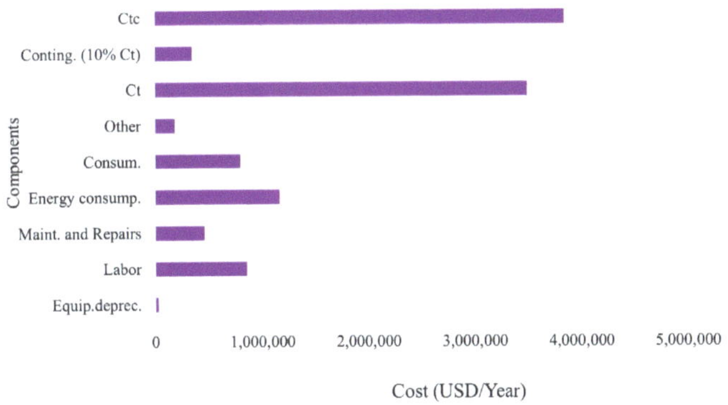

Figure 12. Crushing operating cost components distribution.

Energy consumption, labor, and consumables are the primary drivers of the estimated crushing operating cost, contributing 30%, 22%, and 21%, respectively. Maintenance and repairs, along with equipment depreciation, make up smaller portions, accounting for approximately 12% and 1% of the total cost.

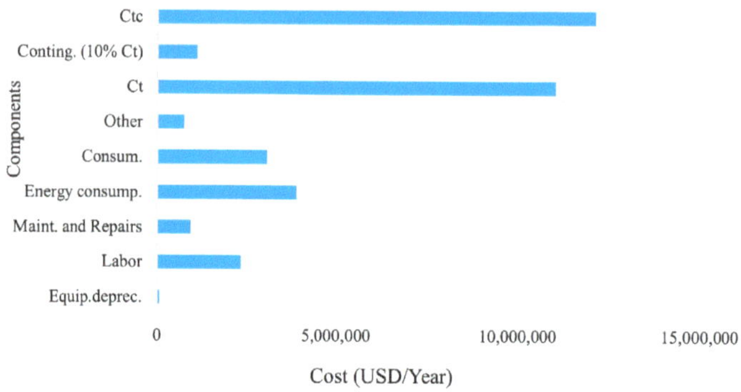

Figure 13. Milling operating cost components distribution.

Milling operating cost is primarily driven by energy consumption, which accounts for 32%, followed by consumables at 25% and labor at 19%. Maintenance and repairs, along with equipment depreciation, represent smaller shares, contributing around 8% and 1%, respectively. An additional 6% falls under the other category, capturing miscellaneous expenses essential to the overall milling process.

3.4. Operating Costs Estimation: Alternative Designs

3.4.1. Drilling and Blasting

Tables 15 and 16 show the results of the drilling and blasting operating cost estimate.

Table 15. Drilling operating cost estimation.

Parameter	Baseline	Design 1	Design 2	Design 3	Design 4
Volume/hole (m^3)	170	180	140	105	75
Tons/hole (t)	442	468	364	273	195
Number of drill holes	14,038	14,038	14,038	14,038	14,038
Drill rig utilization (%)	85	85	85	85	85
Material drilled (t/year)	5,274,076.60	5,584,316.40	4,343,357.20	3,257,517.9	2,326,798.5
C_{tc} (USD/year)	1,814,495.18	1,814,495.18	1,814,495.18	1,814,495.18	1,814,495.18
OPEX (USD/t)	0.34	0.32	0.42	0.56	0.78
Variation (%)		−5.88	+23.53	+64.71	+129.41

Table 16. Blasting operating cost estimation.

Parameter	Baseline	Design 1	Design 2	Design 3	Design 4
Charge length (m)	Variable	7	8	9	10
Mass of charge (t/year)	1009.32	1386.87	1584.98	1783.09	1981.21
Explos. and Acces. (USD/year)	4,768,272.93	6,016,823.84	6,671,979.91	7,327,135.98	7,982,292.06
Ore blasted (t/year)	232,300.92	262,791.36	204,393.28	153,294.96	109,496.40
Waste blasted (t/year)	6,084,645.73	6,306,992.64	4,905,438.72	3,679,079.04	2,627,913.60
Material blasted (t/year)	6,316,946.64	6,569,784.00	5,109,832.00	3,832,374.00	2,737,410.00
C_{tc} (USD/year)	6,419,029.45	7,792,435.46	8,513,107.13	9,233,778.81	9,954,450.49
OPEX (USD/t)	1.02	1.19	1.67	2.41	3.64
Increase (%)	-	16.67	63.73	136.27	256.86

With the annual total cost remaining constant, the drilling operating cost rises to 0.78 USD/t, a 129.41% increase over the baseline, as blast parameters, burden, and spacing (Table 2) decrease. This increase is driven by a reduction in the volume per hole across the different designs, leading to a corresponding decline in the total material drilled annually.

It is evident that as the charge length and total mass of explosives increase with adjustments in stemming height and powder factor across the designs (Table 2), the operating cost surges to 3.64 USD/t, a 256.86% increase over the baseline of 1.02 USD/t, reflecting the impact of more intensive blasting operations. This rise is accompanied by a significant increase in annual spending on explosives and accessories, climbing from USD 4,768,272.93 (baseline) to USD 7,982,292.06 in Design 4. As blasting intensity increases and other parameters decrease, the total annual blasted material drops from 6,316,946.64 tons (baseline) to 2,737,410.00 tons. This reduction further escalates the per-ton operating costs.

3.4.2. Comminution

Tables 17–19 summarize the estimated operating costs for the comminution circuit.

Table 17. Crushing operating cost estimation.

		Primary			
Parameter	Baseline	Design 1	Design 2	Design 3	Design 4
E (kWh/t)	-	0.74	0.70	0.66	0.60
Throughput (t/year)	544,438	438,840	513,360	587,880	654,120
E_T (kWh/year)	-	324,741.60	359,352	388,000.8	392,472.00
Energy cost (USD/year)	-	122,882.22	135,978.80	146,819.50	148,511.40
		Secondary			
Parameter	**Baseline**	**Design 1**	**Design 2**	**Design 3**	**Design 4**
E (kWh/t)	-	0.1002	0.1002	0.1002	0.1002
E_T (kWh/year)	-	3517.72	4115.11	4712.41	5243.47
Energy cost (USD/year)	-	1331.11	1557.16	1783.18	1984.13
		Total			
Parameter	**Baseline**	**Design 1**	**Design 2**	**Design 3**	**Design 4**
E (kWh/t)	1.55	0.8402	0.8002	0.7602	0.7002
Throughput (t/year)	544,438	438,840	513,360	587,880	654,120
E_T (kWh/year)	843,878.9	328,259.32	363,467.11	392,713.21	397,715.47
Energy cost (USD/year)	1,167,495.57	124,213.33	137,535.96	148,602.68	150,495.53
C_{tc} (USD)	3,843,930.37	3,166,528.73	3,181,183.62	3,193,357.01	3,195,439.15
OPEX (USD/t)	7.06	7.22	6.20	5.43	4.89

Table 18. Milling operating cost estimation.

Parameter	Baseline	Design 1	Design 2	Design 3	Design 4
E (kWh/t)	17.62	12.91	12.91	12.91	12.91
Throughput (t/year)	544,438	438,840	513,360	587,880	654,120
E_T (kWh/year)	9,592,997.56	5,665,424.40	6,627,477.60	7,589,530.80	8,444,689.20
Energy cost (USD/year)	3,891,651.91	2,143,796.59	2,507,837.52	2,871,878.45	3,195,470.39
C_{tc} (USD/year)	12,131,689.16	10,209,048.31	10,609,493.33	11,009,938.36	12,044,597.66
OPEX (USD/t)	22.28	23.26	20.67	18.73	18.41

Table 19. Comminution operating cost estimation summary.

Parameter	Baseline	Design 1	Design 2	Design 3	Design 4
E (kWh/t)	19.17	13.75	13.71	13.67	13.61
Throughput (t/year)	544,438	438,840	513,360	587,880	654,120
E_T (kWh/year)	10,436,876.46	5,993,683.72	6,990,944.71	7,982,244.01	8,842,404.67
Energy cost (USD/year)	5,059,147.48	2,268,009.92	2,645,373.48	3,020,481.13	3,345,965.92
C_{tc} (USD/year)	15,975,619.53	13,375,577.04	13,790,676.95	14,203,295.37	15,240,036.81
OPEX (USD/t)	29.34	30.48	26.87	24.16	23.30
Variation (%)	-	+3.9	−8.4	−17.7	−20.6

The results revealed a distinct trade-off between throughput, total energy consumption (E_T), and energy costs across the various designs. As throughput increases from Design 1 to Design 4, E_T also escalates, with Design 4 consuming the most energy. Notably, the E_T of these four designs remained comparatively low when contrasted with the baseline. Furthermore, adjustments to blast design parameters lead to a reduction in the crushing operating cost from 7.06 USD/t to 4.89 USD/t.

A similar trend is observed in milling; as throughput increases, E_T also rises. However, the E_T for these four alternative designs remained significantly lower than that of the baseline. In contrast, the implementation of new blast design parameters significantly reduces milling operating costs from 22.28 USD/t to 18.41 USD/t.

The proposed blast designs, characterized by more intensive blasting, demonstrated a clear link between improved fragmentation, higher throughput, reduced specific energy consumption, and lower comminution operating costs. As finer fragmentation reduces the workload on crushing and milling, the circuit processes material more efficiently, resulting in higher throughput and lower specific energy consumption per ton. This efficiency translates into significant cost reduction, with OPEX dropping from 29.34 USD/t to 23.30 USD/t, a 20.6% reduction in the most optimized design. Although Design 1 shows a slight 3.9% increase in OPEX, subsequent designs achieve significant cost reductions, highlighting how optimized blast parameters improve comminution circuit performance and lower operational costs.

3.5. Summary of Operating Costs Estimation

Table 20 and Figures 14 and 15 present a summary of the estimated operating costs for drilling, blasting, and comminution operations.

Table 20. Operating costs estimation summary.

Operation	Baseline	Design 1	Design 2	Design 3	Design 4
Drilling (USD/t)	0.34	0.32	0.42	0.56	0.78
Blasting (USD/t)	1.02	1.19	1.67	2.41	3.64
Crushing (USD/t)	7.06	7.22	6.20	5.43	4.89
Milling (USD/t)	22.28	23.26	20.67	18.73	18.41
Total OPEX (USD/t)	30.7	31.99	28.96	27.13	27.72
Variation (%)	-	+4	−6	−12	−10

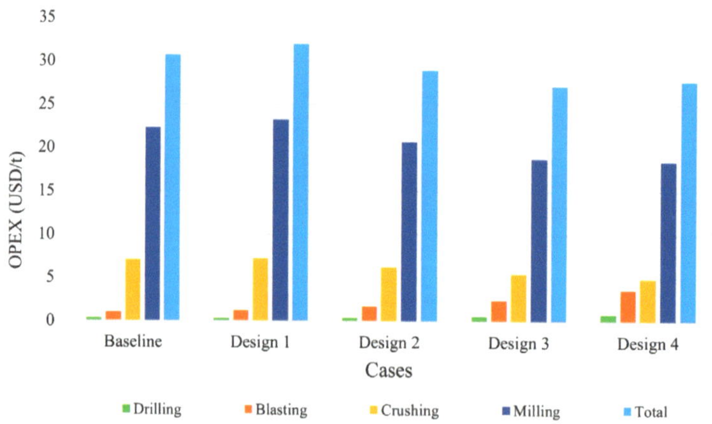

Figure 14. Operating costs estimation results.

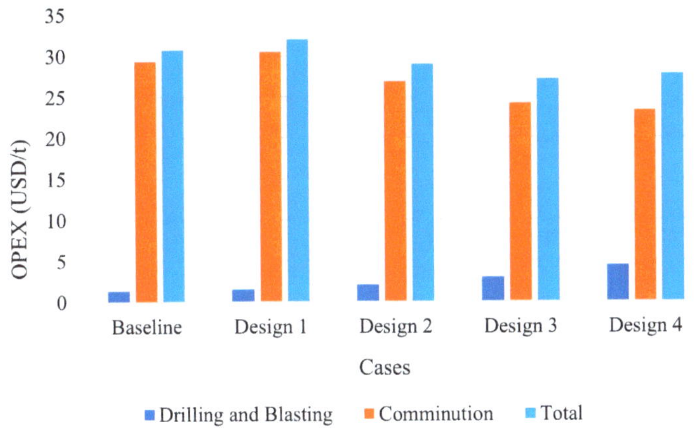

Figure 15. Drilling and blasting versus Comminution operating costs.

The analysis highlights how changes in blast design influence operating costs across drilling, blasting, crushing, and milling, ultimately shaping total OPEX. As blast intensity increases from Design 1 to Design 4, drilling costs rise from 0.34 USD/t to 0.78 USD/t, and blasting costs from 1.02 USD/t to 3.64 USD/t of operating costs. However, these higher upfront expenditures are compensated by reductions in downstream comminution costs, with crushing costs decreasing from 7.06 USD/t to 4.89 USD/t and milling costs from 22.28 USD/t to 18.41 USD/t, driven by improved fragmentation.

This demonstrates how finer particle sizes achieved through more aggressive blasting reduce the energy and effort needed for crushing and milling. While Design 1 increases total OPEX by up to 31.99 USD/t, designs 3 and 4 achieve reductions by up to 27.13 USD/t and 27.72 USD/t, respectively, compared to the baseline. This corresponds to a 4% increase in total OPEX for Design 1 and reductions of 12% and 10% for designs 3 and 4, respectively, showing the trade-off between higher blasting costs and comminution savings. As shown in Figures 14 and 15, this balance reflects the principles of mine-to-mill optimization, where increased blasting efforts improve downstream efficiency, resulting in reduced operating costs across operations [16,22].

While the results demonstrated the benefits of optimizing blast design to enhance comminution circuit performance and reduce operating costs, several limitations must be acknowledged, such as the following:

- Variability in geological conditions: The study assumed relatively consistent geological conditions throughout the blasting and comminution processes. However, natural variability in ore hardness, rock structure, and fragmentation behavior can significantly influence both blast performance and downstream processes. Variations in geological conditions could lead to inconsistent fragmentation, requiring additional adjustments in blasting or processing parameters that were not accounted for in this analysis;
- Predictive model constraints: The Kuz-Ram model, used to estimate fragment size distribution, provides a useful framework but may not capture all the complexities of fragmentation under real mining conditions. In particular, factors like blast-induced damage beyond the intended fragmentation zone or non-linear rock responses could introduce deviations from the predicted performance. This could affect the reliability of the expected comminution savings;
- Balancing economic and environmental considerations: While the study identified cost reductions at the comminution stage, the increased consumption of explosives and higher drilling intensity could raise environmental concerns and impact sustainability metrics. These trade-offs are important when considering operational changes, as the long-term benefits of lower comminution costs may need to be weighed against higher emissions or environmental impacts from intensified blasting.

Future research could explore several key areas, including applying the methodology to various minerals and ores, integrating it with downstream processes beyond comminution, examining the impact of geological variability, incorporating environmental and sustainability considerations, and leveraging advanced analytics and artificial intelligence (AI)-driven approaches for blast optimization.

4. Conclusions

Blasting is the critical initial phase in mining operations, and its impact extends to all subsequent operations. This study assessed the impact of surface blast design parameters on the performance of a comminution circuit processing a copper-bearing ore. Analyzing the effects of varying blast design parameters such as burden, spacing, stemming height, and powder factor on the comminution circuit performance has revealed several key findings.

Firstly, the implementation of new blast patterns and an increased powder factor significantly enhanced run-of-mine (ROM) fragmentation, as evidenced by the notably lower X80 values of the proposed designs compared to the baseline. These modifications to the surface blast design parameters also resulted in substantial improvements in comminution circuit performance, leading to increased throughput and reduced specific energy consumption. Secondly, optimizing energy consumption in comminution presented a valuable opportunity to enhance the mining industry's efficiency by lowering operating costs and minimizing greenhouse gas emissions, which are crucial considerations in the context of global warming and sustainability.

Although this study provided valuable insights for optimizing blast designs to enhance comminution circuit performance, it is important to recognize that these findings are based on theoretical models. Therefore, validating the proposed blast designs with plant data after implementing the optimal blast conditions is essential to confirm that the predicted performance improvements can be achieved under real operational conditions.

Future research is crucial for exploring parameters not addressed in this study, such as the effects of different explosive types and delay timing on blast performance and comminution circuit efficiency. Investigating these factors could provide valuable insights into optimizing blast designs and enhancing operational effectiveness. Understanding how variations in explosive characteristics and timing influence fragmentation and energy consumption may lead to further innovations, ultimately contributing to more sustainable mining practices.

Author Contributions: Conceptualization, H.L.M., B.H. and F.K.M.; methodology, H.L.M., B.H. and F.K.M.; validation, B.H., F.K.M. and J.K.K.K.; formal analysis, H.L.M., P.M.F. and J.K.K.K.; investigation, H.L.M. and F.K.M.; resources, H.L.M., B.H. and P.M.F.; writing—original draft preparation, H.L.M.; writing—review and editing, H.L.M., B.H., F.K.M., P.M.F. and J.K.K.K.; visualization, H.L.M. and P.M.F.; supervision, B.H., F.K.M. and J.K.K.K.; project administration, B.H. and P.M.F. All authors have read and agreed to the published version of the manuscript.

Funding: This work was supported by the National Natural Science Foundation of China (52174185).

Data Availability Statement: The original contributions presented in the study are included in the article; further inquiries can be directed to the corresponding author.

Acknowledgments: The authors express their gratitude to the anonymous mining company for their collaboration and cooperation during this study.

Conflicts of Interest: The authors declare no conflicts of interest.

References

1. Yu, Z.; Shi, X.; Zhou, J.; Chen, X.; Qiu, X. Effective Assessment of Blast-Induced Ground Vibration Using an Optimized Random Forest Model Based on a Harris Hawks Optimization Algorithm. *Appl. Sci.* **2020**, *10*, 1403. [CrossRef]
2. Bayat, P.; Monjezi, M.; Mehrdanesh, A.; Khandelwal, M. Blasting pattern optimization using gene expression programming and grasshopper optimization algorithm to minimise blast-induced ground vibrations. *Eng. Comput.* **2022**, *38*, 3341–3350. [CrossRef]

3. Abbaspour, H.; Drebenstedt, C.; Badroddin, M.; Maghaminik, A. Optimized design of drilling and blasting operations in open pit mines under technical and economic uncertainties by system dynamic modeling. *Int. J. Min. Sci. Technol.* **2018**, *28*, 839–848. [CrossRef]
4. Hakami, A.; Mansouri, H.; Ebrahimi, F.M.; Dehgahn, M.; Faramarzi, F. Study of the effect of blast pattern design on autogenous and semi-autogenous mill throughput at Gol-e-Gohar iron ore mine. In Proceedings of the 11th International Symposium on Rock Fragmentation by Blasting, Sydney, Australia, 24–25 August 2015.
5. Jethro, M.A.; Ajayi, O.D.; Elijah, O.P. Rock Fragmentation Prediction using Kuz-Ram Model. *J. Environ. Earth Sci.* **2016**, *6*, 110–115.
6. Jug, J.; Strelec, S.; Gazdek, M.; Kavur, B. Fragment Size Distribution of Blasted Rock Mass. *IOP Conf. Ser. Earth Environ. Sci.* **2017**, *95*, 042013. [CrossRef]
7. Amoako, R.; Jha, A.; Zhong, S. Rock Fragmentation Prediction Using an Artificial Neural Network and Support Vector Regression Hybrid Approach. *Mining* **2022**, *2*, 233–247. [CrossRef]
8. Zhang, Z.X.; Sanchidrián, J.A.; Ouchterlony, F.; Luukkanen, S. Reduction of Fragment Size from Mining to Mineral Processing: A Review. *Rock. Mech. Rock. Eng.* **2023**, *56*, 747–778. [CrossRef]
9. Napier, M.T. Is progress in energy-efficient comminution doomed? *Miner. Eng.* **2015**, *73*, 1–6. [CrossRef]
10. Holmberg, K.; Kivikytö-Reponen, P.; Härkisaari, P.; Valtonen, K.; Erdemir, A. Global energy consumption due to friction and wear in the mining industry. *Tribol. Int.* **2017**, *115*, 116–139. [CrossRef]
11. Lois-Morales, P.; Evans, C.; Weatherley, D. Analysis of the size–dependency of relevant mineralogical and textural characteristics to particles strength. *Miner. Eng.* **2022**, *184*, 107572. [CrossRef]
12. Hosseini, S.; Mousavi, A.; Monjezi, M.; Khandelwal, M. Mine-to-crusher policy: Planning of mine blasting patterns for environmentally friendly and optimum fragmentation using Monte Carlo simulation-based multi-objective grey wolf optimization approach. *Resour. Policy* **2022**, *79*, 103087. [CrossRef]
13. Whittle, D.; Wooller, R. Maximising the economic performance of your comminution circuit through cut-off optimization. *Int. J. Surf. Min. Reclam. Environ.* **1999**, *13*, 147–153. [CrossRef]
14. González, J.J.F. The Impact of Classification Efficiency on Comminution Performance and Flotation. Ph.D. Thesis, The University of Queensland, Brisbane, Australia, 2020.
15. Karageorgos, J.; Skrypniuk, J.; Valery, W.; Ovens, G. SAG milling at the Fimiston plant (KCGM). In Proceedings of the International Conference on Autogenous and Semiautogenous Grinding Technology (SAG), Vancouver, BC, Canada, 30 September–3 October 2001.
16. Mikee, D.J. *Understanding Mine to Mill*; The Cooperative Research Centre for Optimising Resource Extraction (CRC ORE), Sir James Foots Building, University of Queensland: Brisbane, Australia, 2013; pp. 5–13.
17. Atasoy, Y.; Valery, W.; Skalski, A. Primary Versus Secondary Crushing at St. Ives (WMC) Sag Mill Circuit. In Proceedings of the International Conference on Autogeneous and Semiautogenous Grinding Technology (SAG), Vancouver, BC, Canada, 30 September–3 October 2001.
18. Hart, S.; Valery, W.; Clements, B.; Reed, M.; Song, M.; Dunne, R. Optimisation of the Cadia Hill SAG Mill Circuit. In Proceedings of the International Conference on Autogenous and Semiautogenous Grinding Technology (SAG), Vancouver, BC, Canada, 30 September–3 October 2001.
19. Valery, W.; Jankovic, A.; Larosa, D.; Dance, A.; Esen, S.; Colacioppo, J. Process integration and optimization from mine-to-mill. In Proceedings of the International Seminar on Mineral Processing Technology, Brisbane, Australia, 29–31 January 2007.
20. Nageshwaraniyer, S.S.; Kim, K.; Son, Y.J. A mine-to-mill economic analysis model and spectral imaging-based tracking system for a copper mine. *J. S. Afr. Inst. Min. Metall.* **2018**, *118*, 7–14. [CrossRef]
21. Dragano, M.A.; Seccatore, J.; Cardu, M.; Marin, T.; Bettencourt, J. Influence of blasting charges and delays on the energy consumption of mechanical crushing. *REM Int. Eng. J.* **2019**, *72*, 345–352. [CrossRef]
22. Howe, S.; Pan, J. Application of Enterprise Optimisation Considering Ultra High-Intensity Blasting Strategies. In Proceedings of the Complex Orebodies Conference, Brisbane, Australia, 19–21 November 2018.
23. Valery, W.; Morrell, S.; Kojovic, T.; Kanchibhotla, S.S.; Thornton, D. Modelling and Simulation Techniques Applied for Optimisation of Mine to Mill Operations and Case Studies. In Proceedings of the VI Southern Hemisphere Conference on Minerals Technology, Rio de Janeiro, Brazil, 10–14 September 2001.
24. Lam, M.; Jankovic, A.; Valery, W.; Kanchibhotla, S. Maximising SAG mill throughput at Porgera gold mine by optimizing blast fragmentation. In Proceedings of the International Conference on Autogenous and Semiautogenous Grinding Technology (SAG), Vancouver, BC, Canada, 30 September–3 October 2001.
25. Ozdemir, K. Evaluation of blast fragmentation effects on jaw crusher throughput. *Arab. J. Geosci.* **2021**, *14*, 2036. [CrossRef]
26. Kuznetsov, V. The mean diameter of fragments formed by blasting rock. *Sov. Min Sci.* **1973**, *9*, 144–148. [CrossRef]
27. Cunningham, C.V.B. The Kuz–Ram model for prediction of fragmentation from blasting. In Proceedings of the First International Symposium on Rock Fragmentation by Blasting, Luleå, Sweden, 23–26 August 1983.
28. Vesilind, A.P. The rosin-rammler particle size distribution. *Resour. Recover. Conserv.* **1980**, *5*, 275–277. [CrossRef]
29. Cunningham, C.V.B. Fragmentation estimations and the Kuz–Ram model—Four years on. In Proceedings of the Second International Symposium on Rock Fragmentation by Blasting, Keystone, CO, USA, 23–26 August 1987.
30. Cunningham, C.V.B. The Kuz-Ram fragmentation model–20 years on. In Proceedings of the Brighton Conference, Brighton, UK, 7–11 September 2005.

31. Gheibie, S.; Aghababaei, H.; Hoseinie, S.H.; Pourrahimian, Y. Modified Kuz-Ram fragmentation model and its use at the Sungun Copper Mine. *Int. J. Rock. Mech. Min. Sci.* **2009**, *46*, 967–973. [CrossRef]
32. Figueiredo, J.; Torres, V.; Cruz, R.; Moreira, D. Blasting Fragmentation Study Using 3D Image Analysis of a Hard Rock Mine. *Appl. Sci.* **2023**, *13*, 7090. [CrossRef]
33. Bond, F. The third theory of comminution. *Trans. Am. Inst. Min. Metall. Pet. Eng. Inc.* **1952**, *193*, 484–494.
34. Bond, F. Crushing & Grinding Calculations Part 1. *Br. Chem. Eng.* **1977**, *2*, 378–385.
35. Rosa, D.; Caron, K.; Valery, W.; Diaz, R.; Mamani, H. Blast Fragmentation Impacts on Downstream Processing at Goldfields Cerro Corona. In Proceedings of the 11th International Symposium on Rock Fragmentation by Blasting, Sydney, Australia, 24–26 August 2015.
36. Esen, S. Fragmentation modelling and the effects of ROM fragmentation on comminution circuits. In Proceedings of the 23rd International Mining Congress and Exhibition of Turkey, Antalya, Turkey, 16–19 April 2013.
37. Shehu, S.A.; Yusuf, K.O.; Hashim, M.H.M. Comparative study of WipFrag image analysis and Kuz-Ram empirical model in granite aggregate quarry and their application for blast fragmentation rating. *Geomech. Geoengin.* **2022**, *17*, 197–205. [CrossRef]
38. Gupta, V.K.; Hussain, M. Estimation of the Specific Energy Requirement for Size Reduction of Solids in Ball Mills. *Trans. Indian Inst. Met.* **2017**, *70*, 303–311. [CrossRef]

minerals

MDPI

Article

Evaluation of Ni-Cu Ore from Zapolyarnoe Based on Mineralogical and Physical Properties before and after Comminution

Arturo H. Bravo *, Holger Lieberwirth and Oleg Popov

Institute of Mineral Processing Machines and Recycling Systems Technology, Technical University Bergakamie Freiberg (TUBAF), Lampadiusstrasse 4, 09599 Freiberg, Germany; holger.lieberwirth@iart.tu-freiberg.de (H.L.); oleg.popov@iart.tu-freiberg.de (O.P.)

* Correspondence: bravo@iart.tu-freiberg.de

Abstract: For the effective comminution and subsequent enrichment of mineral ores, comprehensive knowledge of their mineralogical and physical properties is required. Using an integrated methodology, this study evaluated samples of polymetallic Ni-Cu ore from Zapolyarnoe, Russia. Several analytical techniques were utilised, including optical microscopy, microindentation with Vickers geometry, the Point Load Test, and Mineral Liberation Analysis (MLA). The purpose of this study was to determine mineral associations, physical features, and enrichment during jaw crusher comminution. The acquired properties included the Point Load Strength Index, Vickers Hardness Number, and fracture toughness. The MLA method characterised seven fractions in terms of particle size distribution, degree of liberation, association, and modal mineralogy. Magnetite, pyrrhotite, pentlandite, and chalcopyrite were calculated in terms of wt% and their textural features. The enrichment of each ore phase in fractions with particle sizes smaller than 400 μm was determined. The influence of this enrichment was discovered to be correlated with various textural and structural parameters, such as intergrowth, grain size, and crack morphologies after indentations. In addition, the chromium content of magnetite contributed to an increase in the fracture toughness values. Despite the complexities involved, even limited samples of materials provide valuable insights into processing behaviour, emphasising the importance of considering mineralogical parameters in comminution studies.

Keywords: comminution; ore minerals; copper–nickel deposits; jaw crusher; Point Load Test; Mineral Liberation Analysis; vickers hardness; enrichment

1. Introduction

Manufacturing has recently experienced an increase in the demand for mineral raw materials. To address this increase, the manufacturing has processed ores from less conventional mineral deposits and deposits of lower grades [1–4]. Low-grade ore processing requires new technologies and economic models. In particular, selecting adequate processing technologies is crucial for the economic and sustainable production of metals and non-metals, as the stage of ore processing influences the quality of the final product [5,6].

To process a material, valuable minerals must be liberated, especially considering that comminution is one of the most energy-intensive processes, typically accounting for 30 to 50% of the energy consumption in a mining operation [7]. The field of processing research has technologies for the development of advanced comminution, which have aimed to reduce energy consumption and advance more practical and reliable process models [8]. Still, comminution remains a field for further technological research. For example, obtaining the minimum breakage load for the liberation of valuable minerals, staged recovery, and progressive improvements are approaches that several researchers have targeted for a wide variety of industrial sectors [9–11], e.g., the tracing of minerals from exploration. Selective disintegration and pre-concentration in the crushing stage have collectively been identified as a fundamental approach when discussing comminution and grinding processes [12]. In identifying important parameters for comminution, researchers have studied both the particle size reduction in the feed material as well as the liberation of valuable minerals from the gangue [13].

Selective comminution aims to take advantage of the varying breakage characteristics of minerals within an ore, resulting in liberation at comparably coarse particle sizes. This selectivity is also observed in natural processes, such as rock weathering. Despite some adaptation of processing methods to exploit mineral responses, there is currently no systematic approach to optimise processing for specific ores due to limited knowledge of response behaviours and the technological limitations of the available technologies [14].

Selective fragmentation often occurs at mineral interfaces when they are subjected to stresses beyond their tensile strength or by shear. This breakage mechanism is also known as intergranular breakage, as opposed to intragranular or preferential breakage, and occurs due to the strength differences of mineral grains. Several studies have discussed the dynamics of fragmentation after exposure to impacts or compressive stresses [15–22]. King [16] investigated preferential breakage in grinding and identified six distinct phenomena that occur between random fracture and detachment. These phenomena involve the selective breakage of intergranular and intragranular bonds, as well as the preferential comminution of mineral phases. More recently, it has been reported that breakage is influenced by mineral properties such as hardness, cleavage, and grain size [23–27]. Those studies have discussed the dynamics of fragmentation after exposure to impact or compressive stresses and the importance of the distinction between random and non-random breakage. It is observed that breakage, generated by external forces, is associated with zones of weakness and depends significantly on the intrinsic properties (e.g., shape, size, and strength) of the mineral phases constituting the rock, and non-random breakage has a significant impact on the particle size distribution and liberation characteristics. Thus, comprehending the behaviour of minerals and the nature of their boundaries is crucial in determining the success of the process of comminution. Typically, in the early stages of designing the mineral processing facilities for a new mine, only a small amount of sample material is available, usually from drill cores. This research shows that even from such a small amount of sample material quite comprehensive information on processing behaviour can be drawn, even for such complex questions as selectivity in comminution.

This research assessed the relationship between mineralogical and physical properties and the results of selective comminution in crushing using copper–nickel samples from Zapolyarnoe, one of Russia's largest polymetallic deposits. Zapolyarnoe is part of the Norilsk deposit, which is known for its significant concentrations of platinum group elements (PGEs) [28] and is located on the north-western edge of the Siberian platform in the Russian Arctic and hosts massive amounts of Ni-Cu ore [29]. The geodynamic conditions of this area were controlled by rifting and re-activation of the intra-continental basins [30]. Some authors describe an association with Permian–Triassic flood basalts and intrusion by basaltic magma during the Lower Permian–Triassic, forming a series of gabbroic rocks and leucocratic gabbro of the upper gabbro series of Norilsk deposits [28]. The classical model for Ni-Cu-PGEs states that mineralisation in this area is localised in the Norilsk–Kharaelakh valley and the northwestern flank of the Tunguska flood, with differentiated mafic-ultramafic intrusions along synclines [31–33].

The investigation included the characterisation of the material via optical microscopy of representative samples of both thin and polished sections. Using microindentation Vickers Hardness testing, the microhardness properties were determined for all mineral phases. The material underwent processing at the facilities of the Institute for Minerals Processing Machines and Recycling Systems Technology (IART), located at Technische Universität Bergakademie Freiberg, utilising a jaw crusher. Sub-samples of seven size fractions were subsequently subjected to Mineral Liberation Analysis (MLA) to ascertain the distribution of particle sizes, modal mineralogy, associations, and liberation of minerals. Selectivity and enrichment should at least be investigated in terms of their technical feasibility, focusing on the chalcopyrite, pentlandite, pyrrhotite, and magnetite phases. An integrated evaluation of the physical and mineralogical properties obtained by the Point Load Test, Vickers microindentation, and detailed descriptions of the mineralogy under the microscope was considered appropriate to assess the selective comminution linked to the outcome.

During the investigation, the following questions were discussed:

- Is it possible, based on textural information obtained via optical microscopy, to identify certain mineralogical patterns that can provide valuable information for predicting the liberation of mineral phases of interest?
- Do the strength properties of the rock minerals (olivine and feldspar) and the ore phases differ sufficiently such that pentlandite, chalcopyrite, or pyrrhotite can be stressed by using suitable comminution processes so that the material can be broken down into different particle size classes via preferential breakage to subsequently allow separation with minimal loss of valuable material?
- Is it possible to show patterns in grain boundaries and their behaviour, such as a qualitative observation of interfacial fracture formation or preferential breakage? How does this relate to the observed mineralogical characteristics?
- What can be concluded regarding the data obtained via MLA and the fractions generated with compression stress using jaw crushers?

2. Materials and Methods

2.1. Materials

Ni-Cu Drill core samples were collected from the Zapolyarnoe deposit (Figure 1a). The combined drill cores had a total starting mass of 2 kg. Following a macroscopic observation after obtaining a fresh surface, two samples are selected, from each of which three polished sections and three thin sections that are oriented orthogonally were prepared (Figure 1b,c). This assessment enabled the acquisition of relevant and representative information concerning the material received from St. Petersburg Mining University, despite limited details regarding the location and spatial variation of samples. The drill cores have an inside diameter ranging from 35 mm (mostly) to 48 mm. The samples were evaluated together because, after investigation under a microscope, no significant differences were found in the samples.

(a) (b) (c)

Figure 1. Zapolyarnoe Ni-Cu drill cores. (**a**) Drill core collected from Zapolyarnoe. (**b**) One of the selected samples with orientation marksand (**c**) the respective set of thin and polished sections.

2.2. Methods

The Ni-Cu ore material was characterised using various methods, starting with the determination of the strength index using the Point Load Test (PLT). Once the representative samples were selected, the assays were prepared as described in the following flow chart (Figure 2).

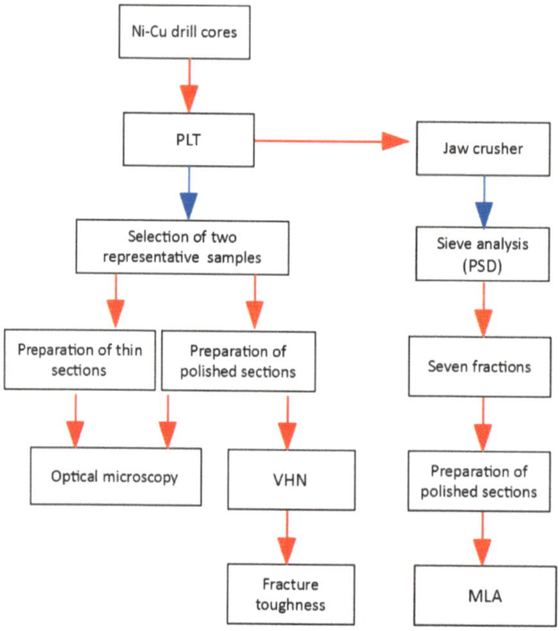

Figure 2. Flowchart of investigations. PLT: Point Load Test. VHN: Vickers Hardness Number. MLA: Mineral Liberation Analysis. PSD: particle size distribution.

2.2.1. Point Load Test

The Point Load Test (PLT) is employed for rock strength classification [34–37]. A notable advantage of this procedure in comparison to other methods (e.g., the Brazilian Test and uniaxial compressive strength) is its limited requirements for the preparation of rock specimens. The test can be conducted quickly and simply with a portable tester either at a quarry or in a laboratory (e.g., Point Load Tester (manufacturer: WILLE Geotechnik)). Fragments of drill cores are sufficient for this method.

The application of a specific load on a rock specimen until breakage occurs defines the strength index (I_s) and can be used for the classification or characterisation of rock samples. The I_s can be calculated from the load on the fracture [36], the platen tip spacing, and the width of the fracture surface (Figure 3). This can be obtained from Equation (1a) using the breaking load (P) (in kN) and the equivalent core diameter (D_e) (in mm), which is obtained from Equation (1b) using the minimum cross-sectional area (A) of the blocky sample. This area is calculated from the distance (D) (in mm) between the contact points and the narrower dimension width W (in mm) of the blocky sample, as in Equation (1c). The incorporation of the equivalent diameter (D_e) is useful to compare results from irregularly shaped specimens with diametral results (Figure 3).

$$I_s = \frac{P}{D_e^2} \quad \text{(1a)}$$

$$D_e^2 = 4\frac{A}{\pi} \quad \text{(1b)}$$

$$A = WD \quad \text{(1c)}$$

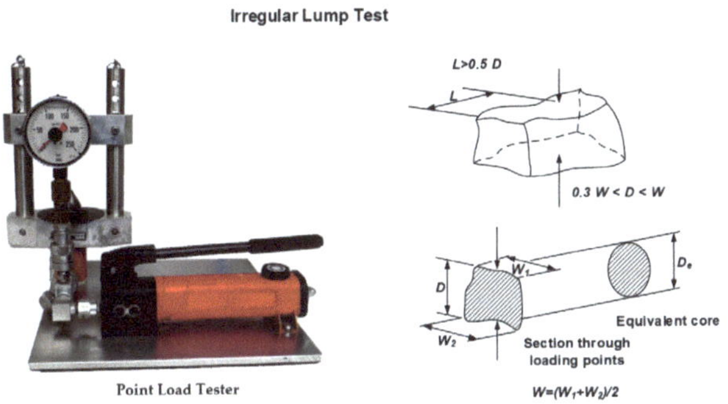

Figure 3. Point Load Tester used (left) with the measurement for irregularly shaped specimens [36].

The I_s depends greatly on the total size of the specimen. This effect, which is also seen with all other procedures (e.g., the uniaxial compression test), is generally referred to in rock mechanics as the scale effect [38]. Since the specimen size influences the strength index, the strength indices of rock pieces of different dimensions must be converted into values for a standard specimen with a standard diameter (D_e = 50 mm) to allow the comparison of results. The strength thereby changes to the $I_{S(50)}$. Broch [34] used an idealised set of curves that should enable the reduction in every individual value for the I_S arbitrary distance (D) to a standard value $I_{S(50)}$. For the possible regression of measurement data on a logarithmic grid [34,36], the I_S specifies a grain size correction factor (f), which enables conversion to the $I_{S(50)}$ for rocks (Equation (2)). In the equation, m is equal to the regression curve's gradient constant.

$$I_{s(50)} = f \cdot \frac{P}{D_e^2} = \frac{P}{D_e^2} \cdot \left(\frac{\mathrm{De}}{50} \right)^{2(1-m)} \tag{2}$$

The strength index was determined using a test instrument from WILLE. The measurement range of this instrument is between 0 and 400 bar. The Point Load Test could be carried out on irregularly shaped particles as well as core samples. The calculation of point load indices (I_S) was performed following the guidelines of ISRM [36]. The $I_{S(50)}$ values were determined according to Raaz [39].

Given the presence of an empirical correlation between the Uniaxial Compressive Strength (UCS) and the Point Load Index ($I_{S(50)}$), it is possible to approximate the order of magnitude of the UCS by multiplying the $I_{S(50)}$ value by a conversion factor (Equation (3)) that requires specific determination for a given ore [40,41]. The relationship between the strength index ($I_{S(50)}$) and the compressive strength (σ_D) as a constant has been described by many researchers [35,37]. The ISRM [36] suggests indicating a relationship for the σ_D and the $I_{S(50)}$ of 20 to 25. On the basis of extensive Point Load tests conducted at the IART, the conversion factor can be defined as c = 25.

$$\sigma_D = c \cdot I_{S(50)} \tag{3}$$

With the limited amounts of sample materials from drill cores, a statistically substantiated result often cannot be obtained, but it is always possible to conduct assays and draw conclusions about the variance in the results. However, quarter and half drill cores, which are often available for the investigations described here, are not sufficient to process regular samples.

For the UCS test, the dimensions still fit for Point Load tests. Furthermore, it helps in the selection of representative samples for subsequent microindentation with Vickers geometry and MLA tests.

2.2.2. Optical Microscopy

Thin and polished sections were prepared after the PLT for a detailed investigation of mineralogical features. The representative samples were examined under an optical microscope with transmitted and reflected light to describe and determine mineralogical data (mineral phases and alterations) and grain boundaries between the different mineral phases. Microscopic analysis was conducted using Zeiss Jena Jenapol (ZEISS, White Plains, NY, USA) and Olympus BX53M polarised-light microscopes equipped with JVC KY-F50 and Olympus U-TV1XC (Olympus, Tokyo, Japan) high-resolution digital cameras, respectively. This analytical method provided a way to identify with a relatively high resolution the phases within the textural framework.

2.2.3. Vickers Hardness Number and Fracture Toughness

While the PLT provides information on macroscopic features such as rock/ore strength, information on the behaviour of the valuable constituents has to be derived from the microscopic properties of the material constituents, the mineral grains, and the grain boundaries. The size of an indentation usually does not exceed 30 μm and this falls well within the size of many mineral grains. Even if a crack propagates towards or along a grain boundary, this may provide valuable information for comminution. The VHN test is seen as a promising approach to obtaining this information, and the morphology of the indentation can be classified.

The hardness of a material is generally expressed in terms of its resistance to local deformation. For its determination, tests based on the material's resistance to indentation are carried out. An indenter is pressed into the surface of the material with a known load. Then, the size of the subsequent indentation is measured: soft materials give a large indentation, and hard materials give a small one. To determine the VHN according to ISO6507 [42], a pyramidal indenter was pressed with an interface angle of 136° on a specimen with a defined test force. For the determination of fracture toughness, the propagation of cracks was considered [42,43]. L_1, L_2, L_3, and L_4 are the lengths of radial cracks originating from four different edges of a Vickers indentation. An ideal case is $C_1 = a_1 + L_1$, $C_2 = a_2 + L_2$, $C_3 = a_3 + L_3$, or $C_4 = a_4 + L_4$ (Figure 4).

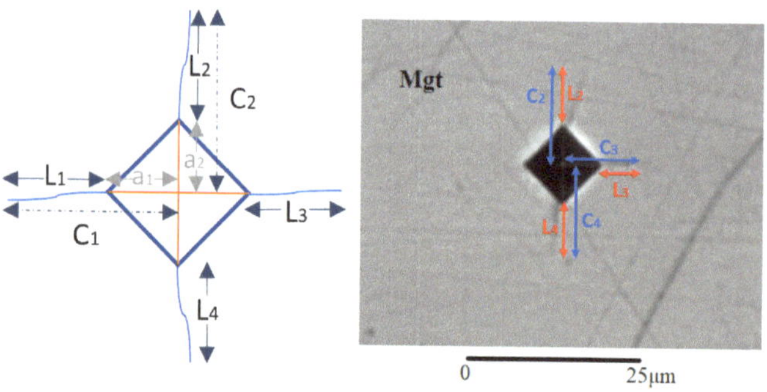

Figure 4. Example of fracture propagation evaluation in magnetite (Mgt). L_1, L_2, L_3, and L_4 are the lengths of the radial cracks.

The Vickers Hardness results from the quotient of the applied test force (F in Newtons (N)) and the surface area of the residual indentation mark on the specimen. To calculate the surface area of the residual pyramidal indentation, the average of the two diagonals (d_1 and d_2 in mm) is used (Equation (4)) because the base areas of Vickers indents are frequently not exactly square (Figure 4). The recommended Vickers Hardness range is given in the standard ISO6507 [42].

$$HV = \frac{F}{A} = 2F \cdot \frac{\sin 68^\circ}{d^2} = \frac{1.8544 \cdot F}{d^2} \quad (4)$$

The crack lengths C_1–C_4 that develop mainly at the corners of the indentation can be used to qualitatively and quantitatively characterise the fracture toughness (Figure 4).

Fracture toughness is measured to account for resistance to crack propagation. Theoretical models for estimating fracture toughness (K_{IC}) fracture toughness are based on a fracture mechanics approach. Previous research has investigated different ways of calculating the mechanical properties corresponding to radial–median cracks and the Palmqvist crack system [44,45]. This investigation considered the model of Blumenauer [46] (Equation (5)) giving values very similar (Equation (6)) [47] to the model of Tanaka [48] (Equation (7)).

$$K_{IC} = \frac{F_i}{(\pi \cdot c_n)^{\frac{3}{2}} \cdot \tan\left(\frac{\alpha}{2}\right)} \tag{5}$$

$$K_{IC} = 0.0726 \cdot \frac{F}{C^{\frac{3}{2}}} \tag{6}$$

$$K_{IC} = 0.0725 \cdot \frac{F}{C^{\frac{3}{2}}} \tag{7}$$

The investigation of the hardness and fracture toughness of different mineral phases via Vickers Hardness testing was carried out using the Shimadzu HMV-G21DT Micro Vickers Hardness tester. To provide a simplified description of the indentation morphologies, this study classified the results into five groups (Figure 5).

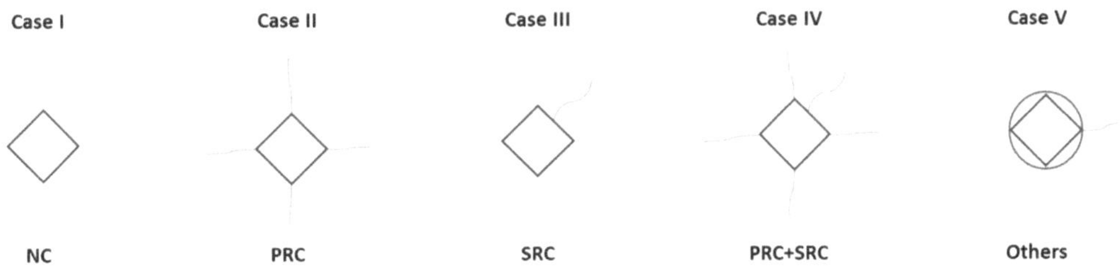

Figure 5. Morphology classification after indentations according to the most typical cases for this study. NC: no cracks; PRC: primary radial crack (which may have one, two, three, or four radial cracks (RCs)); SRC: secondary radial crack (which may have one, two, three, or four radial cracks (RCs)). Case IV includes all possible scenarios where primary and secondary radial cracks occur together. Additionally, case V encompasses other morphologies such as cone cracks or other irregular morphologies that cannot be classified in the previous categories.

The classification considers the formation of primary radial cracks (PRCs) and secondary radial cracks (SRCs), representing case II and case III, respectively. Primary radial cracks have an origin that lies in structures that are initiated after the indentation, and these cracks extend outward from the indentation site towards the edges. Subsequently, secondary radial cracks are mainly formed after primary cracks and propagate around them. The classification also includes cases where primary and secondary radial cracks are combined (case IV) and other types of cracks (case V) that may occur (e.g., if radial cracks are generated). If no cracks (NC) are observed after indentation, it is considered case I (Figure 5).

2.2.4. Mineral Liberation Analysis (MLA)

MLA can provide information on the particle shape and particle size distribution of smaller progenies and their modal mineralogy, mineral associations, and liberation. It is an important tool to evaluate the result of a comminution process, for instance with regard to selectivity. MLA combines an automated Scanning Electron Microscope (SEM) with multiple energy-dispersive X-ray detectors (EDS) to automatically acquire and jointly analyse images and elemental spectral information. SEM analysis of the samples enables automated analyses based on the evaluation of backscattered electron (BSE) images and the acquisition of energy-dispersive X-ray spectra of particles visible in the BSE imaging mode [49–51].

MLA analysis of Ni-Cu ore samples was performed at the Geometallurgy Laboratory, Institute of Mineralogy, TUBAF, using a scanning electron microscope FEI quanta 600 FEG (FEI, Hillsboro, OR, USA) equipped with a field emission source, two Bruker energy dispersive X-ray (EDX) SDD detectors, Bruker Quantax 200 with two Dual Xflash 5030 EDX detectors (Bruker, Berlin, Germany) and backscattered electron (BSE). The greyscale of the BSE image was calibrated with epoxy resin as the background (BSE grey level) and gold (pin in RDI standard block) as the upper limit (BSE grey level of about 250). The EDS X-ray spectrometers were calibrated with copper (pin in the RDI standard block). Several measurement modes are available in the MLA expert software, each calibrated for specific applications. In this study, the measurement mode GXMAP, i.e., grain X-ray mapping, was used. GXMAP uses X-ray mapping to identify phases that cannot be segmented using BSE grey levels alone. The Ni-Cu ore, which was previously crushed and divided into seven fractions, was analysed using MLA 2.9 software package (JKTech, Brisbane, Australia). For this purpose, the samples were embedded in epoxy resin. The ground surfaces of the resulting sections were coated with carbon and examined via SEM-EDS to identify the chemical composition and texture of the samples.

The mineral locking and degree of liberation data for each phase were obtained through MLA analysis. The mineral locking serves as an indicator to determine if a phase of interest is liberated or associated solely with another. The degree of liberation indicates the percentage of the ore's total content that consists of mineral particles that are not bound [52]. It is possible to represent and calculate it in one of the following ways:

Surface area: Liberation of a particle, expressed as a percentage, denotes the length fraction on the outer perimeter occupied by the mineral or minerals of interest in relation to the entire outer perimeter of the particle.

Volume: The area fraction of the mineral or minerals of interest in relation to the total area of the particle, typically expressed as a percentage, is denoted by the volume of liberation of the particle.

The degree of liberation of a mineral is frequently calculated in process mineralogy through the examination of two-dimensional sections of a particle set that is statistically representative and contains the mineral of interest. It is usually assessed based on the area percentage of the mineral grain(s) within a particle. Particles are often categorised into various grades based on incremental steps, such as 10% increments (100% liberated, 90–100% liberated, 80–90% liberated, etc.), or broader 30% steps such as <30% (locked), 30–60% (middlings), and >60% (liberated).

3. Results

3.1. Description of Optical Microscopy

Concerning the characteristics of the Zapolyarnoe samples, different mineral phases and relationships between them were identified and described. The results for rock-forming minerals and ore mineral phases are detailed in this section. The textural and structural information obtained is crucial for understanding and improving the size reduction process, mainly with respect to liberation size.

More than ten minerals were identified, mainly of mafic and calcium-rich phases such as olivine, mica (biotitic), and feldspar.

The ore was mainly composed of magnetite, pyrrhotite, pentlandite, and chalcopyrite with accessory amounts of ilmenite. Magnetite is the most abundant primary mineral of the ore and was mainly distributed in contact with olivine, pyrrhotite, ilmenite, and chalcopyrite (Figure 6d–f). Magnetite crystals were generally smaller than 0.4 mm (Figure 6e,f). Pyrrhotite generally occurred as anhedral grains. Grains with 1–2 mm diameters were commonly fractured and contained small euhedral to subhedral pyrrhotite crystals (Figure 6a). Pentlandite occurred as grains of various sizes enclosed within pyrrhotite and chalcopyrite. Pentlandite showed cataclastic fractures infilled with chalcopyrite and pyrrhotite (Figure 6b). Solid-state exsolution may occur when certain high-temperature minerals undergo cooling. In the case of pentlandite, a typical flame-like exsolution was observed in pyrrhotite (Figure 6c). The paragenesis between pentlandite and pyrrhotite occurred in intergrowth between them and was disseminated throughout the sample. Chalcopyrite was also found along the edges of the intergrowth (Figure 6b,c). Ore minerals filled the interstices between crystals of pyroxene, olivine, and feldspar (mainly plagioclase). This means that the copper and nickel sources occurred later relative to the main rock-forming minerals. As for the main phases of the sample, olivine appeared to be the most common phase. Olivine was observed in relics of rhombic crystals (Figure 6d,g,i). Poikilitic pyroxene crystals with rounded olivine inclusions (Figure 6e) and plagioclase laths (Figure 6i) with polysynthetic twins were observed. Locally, the pyroxene formed subidiomorphic and elongated prismatic crystals measuring 1–2 mm (Figure 6h). Brown ferruginous biotite had short laminar and tabular forms ranging from 0.2 to 0.5 mm (Figure 6h) and locally it was possible to identify chlorite.

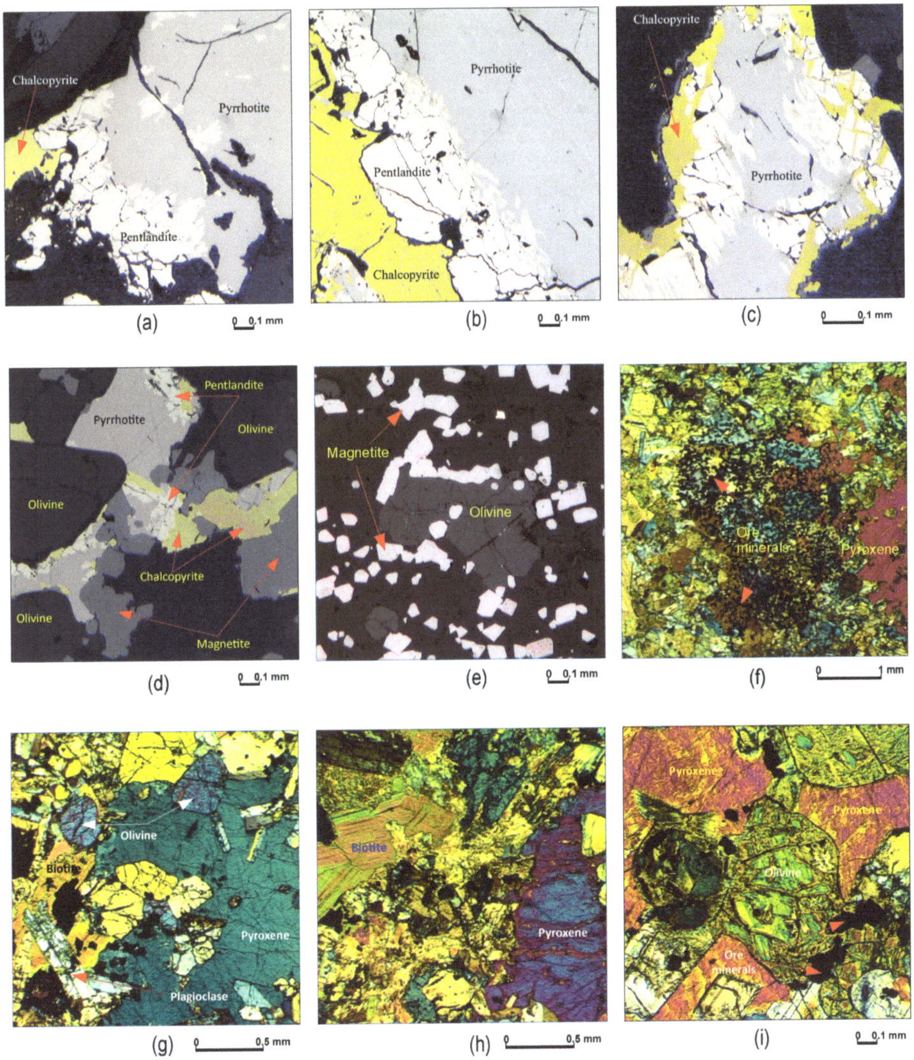

Figure 6. Microphotographs of thin sections from Ni-Cu Zapolyarnoe ore deposit samples. (**a**) Intergrowth of pentlandite, pyrrhotite, and chalcopyrite crystals on the edges. Locally, pentlandite

is an aggregate (**b**) intergrowth between pentlandite, pyrrhotite, and chalcopyrite. The crystals of pentlandite are clearly fractured. (**c**) Chalcopyrite and pentlandite with exsolution texture. Phases are in contact with rock-forming minerals. (**d**) Olivine crystals in contact with ore minerals. Magnetite in contact with pyrrhotite and chalcopyrite. (**e**) Disseminated crystals of magnetite distributed in a scattered form. (**f**) Disseminated magnetite crystals (ore minerals). (**g**) Plagioclase laths with poikilitic texture. Olivine crystals are enclosed by the pyroxene oikocryst. (**h**) Biotite and main mineral phases. (**i**) Poikilitic pyroxene crystal with rounded olivine inclusions and plagioclase laths; iddingsite alteration in olivine is observed.

3.2. Point Load Strength Index

To evaluate the rock strength, 12 specimens were taken from the sampling area to carry out the Point Load Test. This number was relatively small because of the small amount of drill core sample material available. The Point Load Index of the Ni-Cu samples was $I_{S(50)} = 3.94 \pm 1.23$ MPa. In accordance with Equation (3), this corresponds to an average uniaxial rock compressive strength of $\sigma_D = 98.4$ MPa. According to the ISRM strength scale (Figure 7), the tested Zapolyarnoe samples were classified as rocks with high strength. The structural homogeneity, which is the variation in the coefficient in relation to middle size ($\sigma/\overline{x}$), was calculated to be 31.2% for the Zapolyarnoe specimens. This value was high, which meant that the irregular rock samples were heterogeneous and a larger sample would normally be recommended.

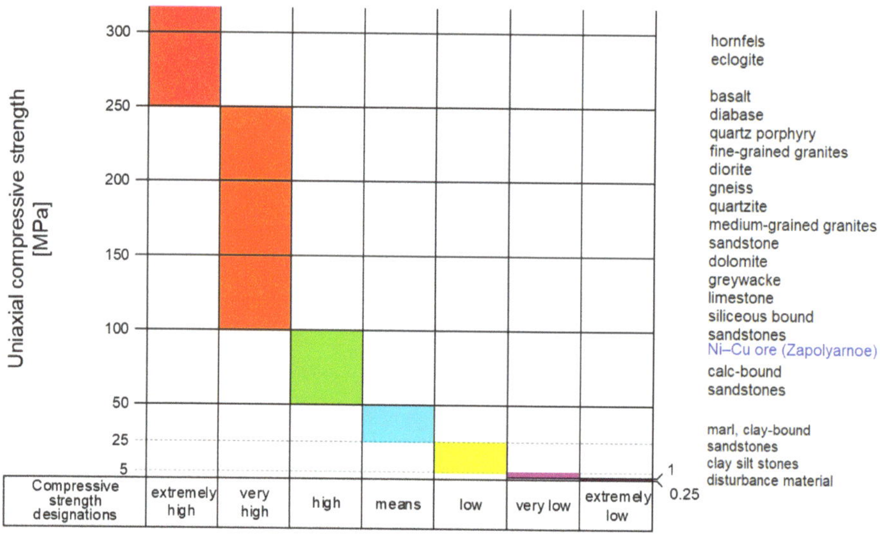

Figure 7. Data from commonly used descriptions for the uniaxial compressive strength of rocks [53] in accordance with the selected sample (Ni-Cu ore) highlighted.

3.3. Microindentations and Crack Morphology Results

Up to 50 microindentations were made for each of the mineral phases. The fracture toughness values and Vickers Hardness Numbers of the mineral phases were determined (Table 1). This study selected areas in the phases that purposely avoided pores and microfractures. The results for the Vickers Hardness (Figure 8) and fracture toughness (Figure 9) were plotted and compared. To avoid influences that could originate from the grain boundaries, these points were analysed separately.

Table 1. Vickers Hardness Number (VHN) in n/mm^2 and fracture toughness (K_{IC}) in MN/m$^{3/2}$ for each main phase.

Mineral Phase	VHN		K_{IC}	
	Mean	Stdev	Mean	Stdev
Pentlandite	254.91	2.95	1.59	0.33
Pyrrhotite	303.65	7.09	0.90	0.14
Chalcopyrite	211.22	7.91	0.63	0.16
Magnetite	866.35	80.84	1.63	0.62
Olivine	968.37	50.32	1.24	0.48
Feldspar	816.76	46.10	1.33	0.40

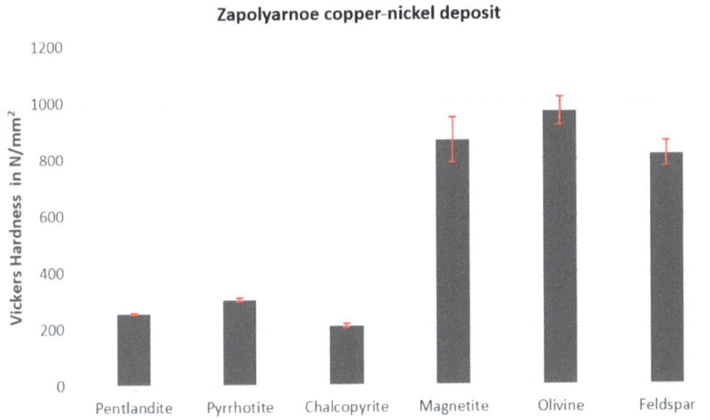

Figure 8. Vickers Hardness Number obtained for the main mineral phases in Zapolyarnoe.

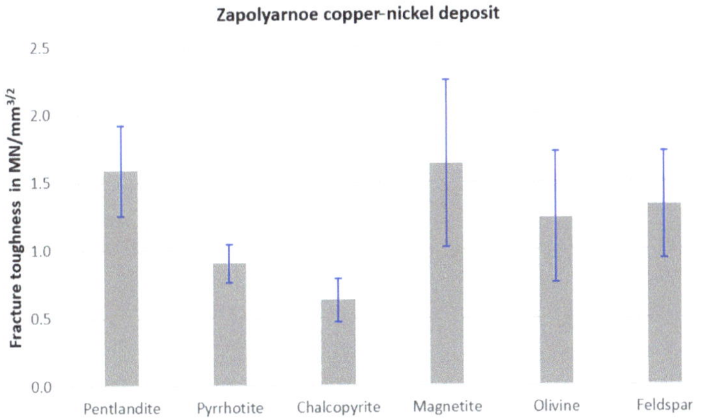

Figure 9. Fracture toughness values obtained for the main mineral phases from Zapolyarnoe.

The microindentations were carried out on a mesh of approximately 20 μm from an initial point close to the grain boundary (Figure 10a). The load was 245.2 mN for chalcopyrite, pentlandite, and pyrrhotite with a holding time of 15 s.

The same time was used for magnetite, olivine, and feldspar, but the load was raised to 490.3 mN. The reason for this is that these minerals have greater hardness, so the indentations resulting from this load were very small at the scale of observation that was used.

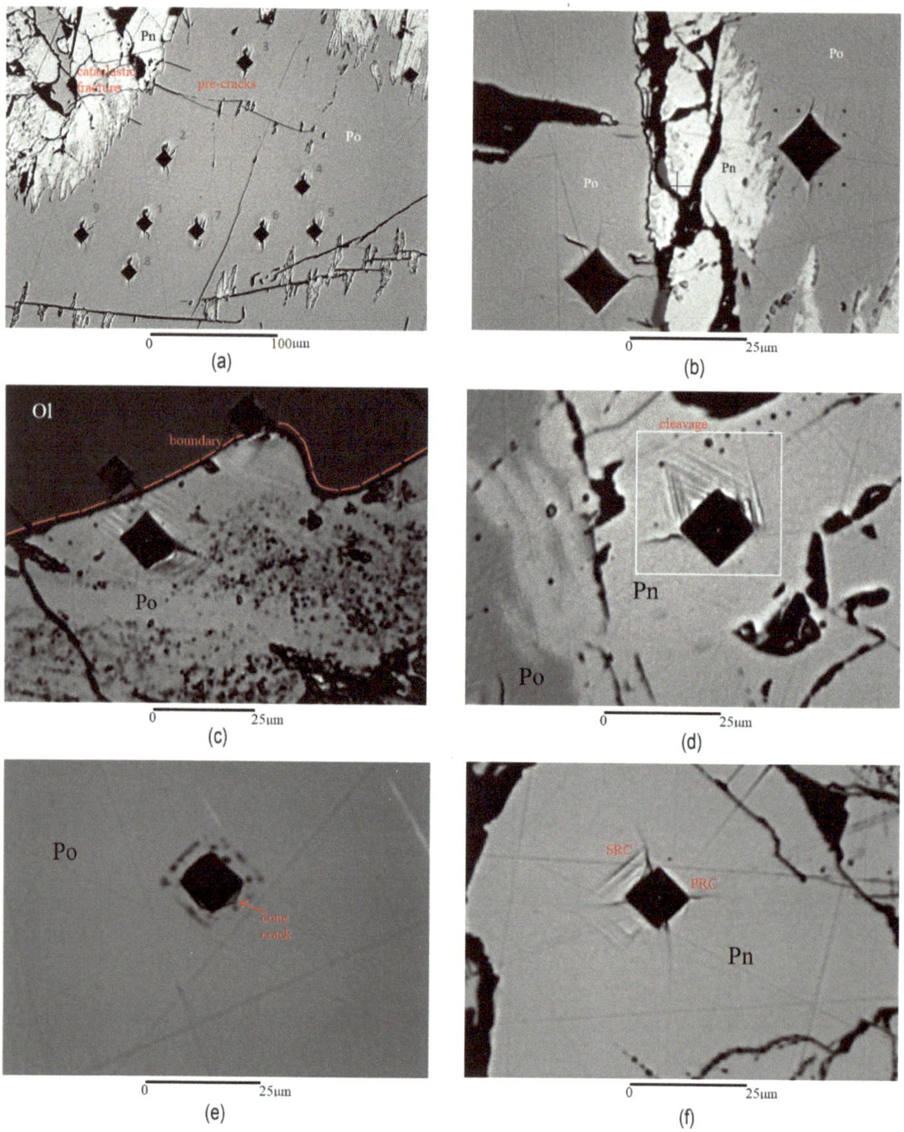

Figure 10. Vickers Hardness microindentation investigation. (**a**) Systematic analysis of of a polished section,considering an equidistant mesh and avoiding irregularities (marks 1 to 9). (**b**) Example of microindentation close to a boundary. (**c**) Boundary between a rock-forming mineral (olivine) and an ore phase (pyrrhotite). (**d**) Lamination after indentation in pentlandite. (**e**) Cone crack observed after indentation in pyrrhotite. (**f**) Case IV according to morphology description (PRC + SRC) in pentlandite. (Pn) pentlandite; (Po) pyrrhotite; (Ol) olivine; (Cp) chalcopyrite.

Upon analysing the Vickers Hardness data, it is evident that the hardness values of ore minerals were often lower than those of rock-forming minerals, except for magnetite, which exhibited both high hardness and high fracture toughness. A significant chromium content (11.72% wt) explained this behaviour after indentation, as often no cracks occurred in magnetite. For pyrrhotite, only small cracks were observed (Figure 10a). Furthermore, a cone crack was detected in pyrrhotite (Figure 10e). In chalcopyrite, crack propagation was evident in two directions (L_2 and L_3), but in the other two directions crack propagation was negligible (Figure 10e). Most of the crack morphologies for chalcopyrite were classified as PRCs. In the case of pentlandite, the mineral phase itself contained original cracks (Figure 10a,b). Owing to the presence of such irregularities in the mineral phase, the crack propagation showed no common pattern (Figure 10f). In the case of magnetite, it was common to observe cracks of the PRC + SRC type.

To provide a clearer visual representation of the predominant types of cracks formed during indentation, a comparative graphical analysis was conducted to examine the occurrence of primary radial cracks, secondary radial cracks, and other possible morphologies (Figure 11). Upon analysis, it was evident that pyrrhotite and chalcopyrite exhibited a generation of primary radial cracks (PRCs) or the absence of crack propagation (NC). Conversely, in pentlandite, the formation of secondary radial cracks (SRCs) was prevalent. Although SRC generation without the presence of PRCs was uncommon, this was attributed to the significant original cracks present in the pentlandite (Figure 10d).

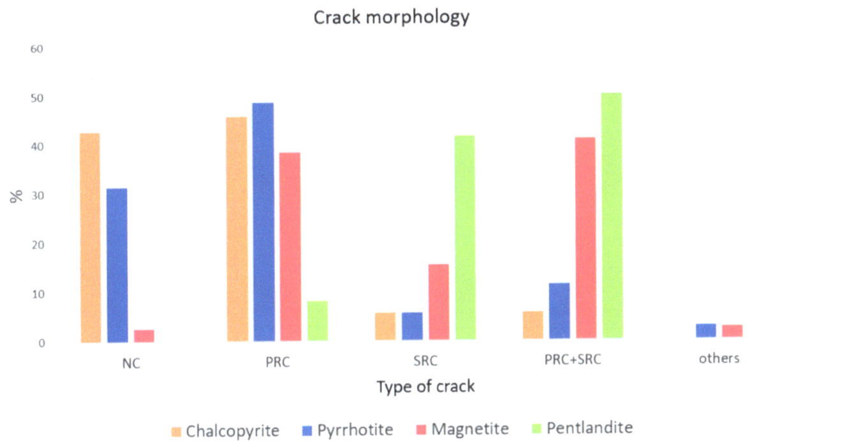

Figure 11. The percentage of cracks according to morphology: primary radial cracks (PRCs), secondary radial cracks (SRC), a combination of both (PRC + SRC), and other types of cracks (others) for the phases of chalcopyrite, pyrrhotite, magnetite, and pentlandite. The graphic also includes the percentage where no cracks (NC) were observed.

For the rock-forming minerals, olivine and feldspar, the microindentation showed a higher hardness value compared to the ore minerals. The response of olivine to microindentation was determined as a trend because primary radial cracks following the L_2 and the L_4 directions were generated in most of the tests. This fracture propagation was oriented towards the boundary (Figure 10c). Delamination textures were also visible in the case of pyrrhotite when indenting near the grain boundary with olivine (Figure 10c). In the case of magnetite studies, cracks were observed in many directions. In a few cases, a radial crack was found near the remaining indentation.

3.4. Mineral Liberation Analysis Results

An experimental study was conducted in the IART Institute facilities using a Single Toggle Jaw Crusher RETSCH model BB 250 XL (Retsch GmbH, Haan, Germany). Size reduction takes place in the wedge-shaped area between a fixed arm and one moved by an eccentric drive shaft. The elliptical motion crushes the sample which is moved towards the crushing gap under gravity. The potential of the jaw crusher was evaluated in terms of pre-concentration/selectivity after material size reduction with a compressive constant input gap of 30 mm.

This model of crusher is used for rapid and effective crushing and pre-crushing of materials. The input material was macroscopically described and weighed, resulting in an initial sample total of 2 kg (the particles after PLT are the feed material for Jaw Crusher). From the crushed product, the particle size distribution was obtained by sieving with apertures sizes of 2800, 1400, 710, 355, 180, and 125 μm. The sieving was performed using a Haver EML 200 sieve shaker (Retsch GmbH, Haan, Germany) for 10 min at a constant amplitude. The obtained fractions were analysed using Mineral Liberation Analysis (Figure 12) and the particle size distribution results of the combined data for all size fractions were plotted (Figure 13). The mass median diameter (D_{50}) was calculated to be 2.93 mm.

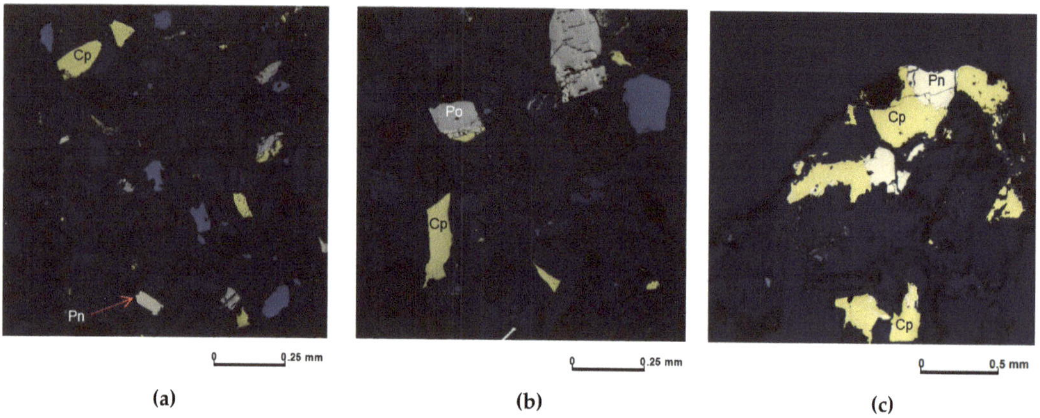

Figure 12. Microphotographs used as examples of liberated ore and non-liberated ore in different fractions. (**a**) Fraction 125/180 μm, (**b**) fraction 180/355 μm, and (**c**) fraction 1400/2800 μm. Cp: Chalcopyrite, Pn: Pentlandite, Po: Pyrrhotite.

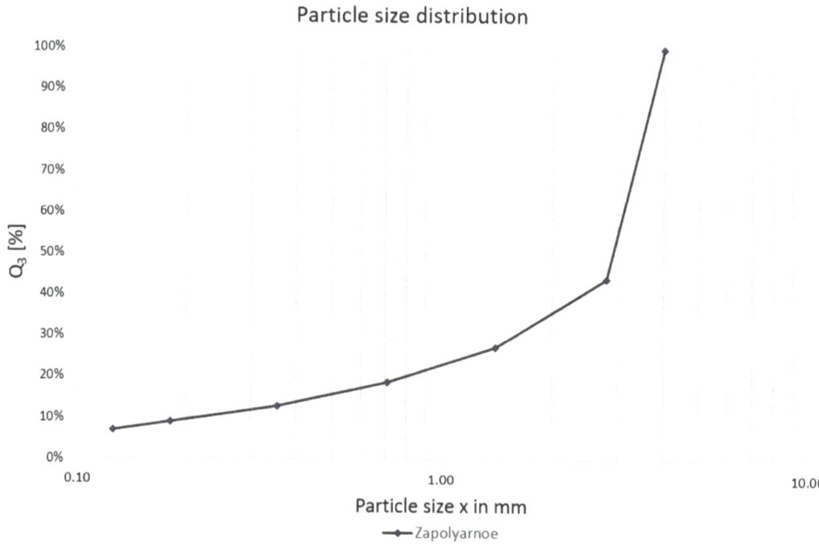

Figure 13. Particle size distribution for the crushed Ni-Cu ore from Zapolyarnoe.

3.4.1. Mode Mineralogy Data

The modal composition of the feed material was determined based on the MLA results. Seventeen phases were identified (Figure 14). The mode mineralogy data obtained for the seven fractions were expressed in wt% for each mineral phase (Figure 15). Optical microscopic analysis of unprocessed materials revealed the prevalence of magnesium-rich forsterite olivine, feldspar, and chromium-rich magnetite phases. The forsterite exhibited its largest concentration in the portion larger than 2800 μm, with 48 wt%.

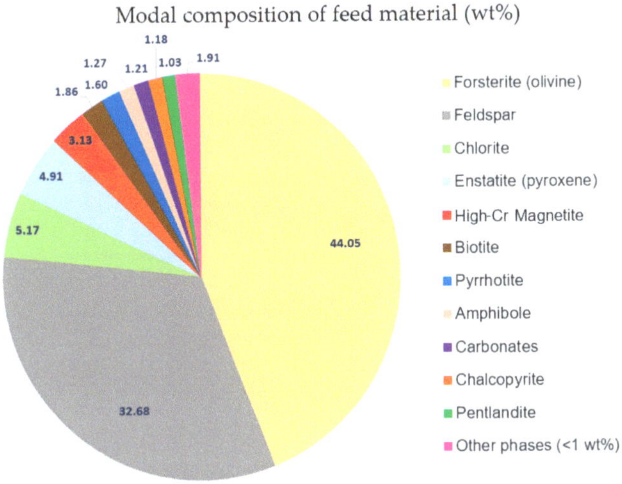

Figure 14. Modal composition (wt%) of the feed material calculated for the Ni-Cu samples. The phases included in the group below 1 wt% correspond to spinel (0.88 wt%), ilmenite (0.33 wt%), quartz (0.33 wt%), pyrite (0.25 wt%), apatite (0.10 wt%), and anhydrite (0.02 wt%).

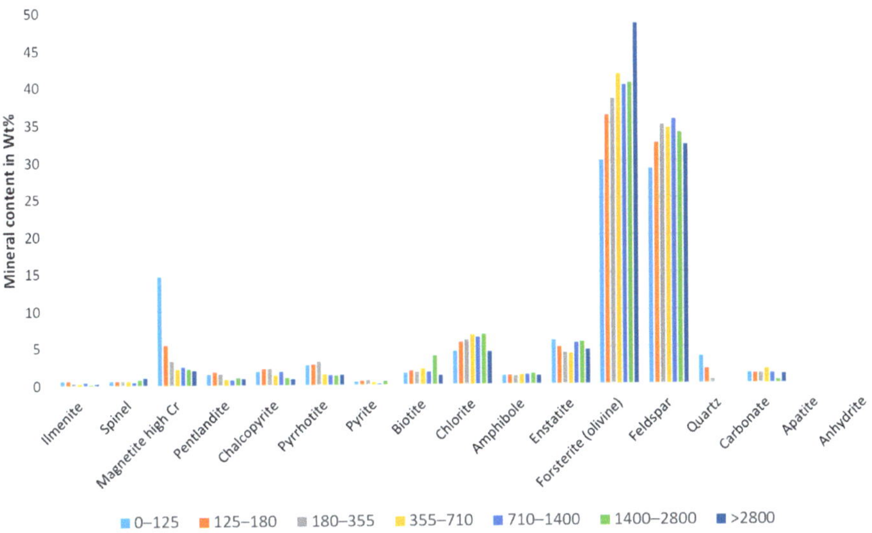

Figure 15. The mineral content of the product as a percentage identified via MLA for each size fraction (in μm) of the Zapolyarnoe samples.

The concentration of feldspar remained reasonably consistent across all fractions, with the variation ranging from 28 to 35 wt%. The results for pyroxene (enstatite), biotite, chlorite (secondary), amphibole, and carbonates remained steady and were consistently below 10 wt% for all fractions. Quartz was exclusively found in small particles. Concerning the ore minerals, magnetite exhibited an increase in weight percentage for smaller particle sizes (less than 125 µm), exceeding 14 wt%. The comparative graph (Figure 15) clearly shows an increase in the amount of chromium-rich magnetite. The abundant presence of magnetite in the smaller fractions was evident based on the mineral sizes identified under the microscope. It typically measured less than 0.4 mm, as previously mentioned in the thin-section descriptions.

3.4.2. Degree of Mineral Liberation

This section presents the data in terms of the liberation degrees obtained for magnetite, pyrrhotite, pentlandite, and chalcopyrite.

The degree of liberation was plotted as a function of the size fraction (in µm). Pentlandite was mainly associated with two or more phases (binary or ternary) in fractions larger than 125 µm, while at fractions up to 355 µm, pentlandite was practically not liberated (Figure 16a). Mineral locking showed that pentlandite was 43.4% liberated (for 0/125 µm particles), 8.7% locked (binary) to pyrrhotite, 3.9% locked (binary) to chalcopyrite, 2.4% locked (binary) to magnetite, and 13.8 % locked (binary) to non-sulphide gangue (NSG). In the case of chalcopyrite, the degree was similar, although with higher liberation in fractions below 355 µm (Figure 16b). It was 50.1% liberated (for 0/125 µm particles), 3.1% locked (binary) to pyrrhotite, 3.7% locked (binary) to pentlandite, 2.6% locked (binary) to magnetite, and 17.2% locked (binary) to non-sulphide gangue.

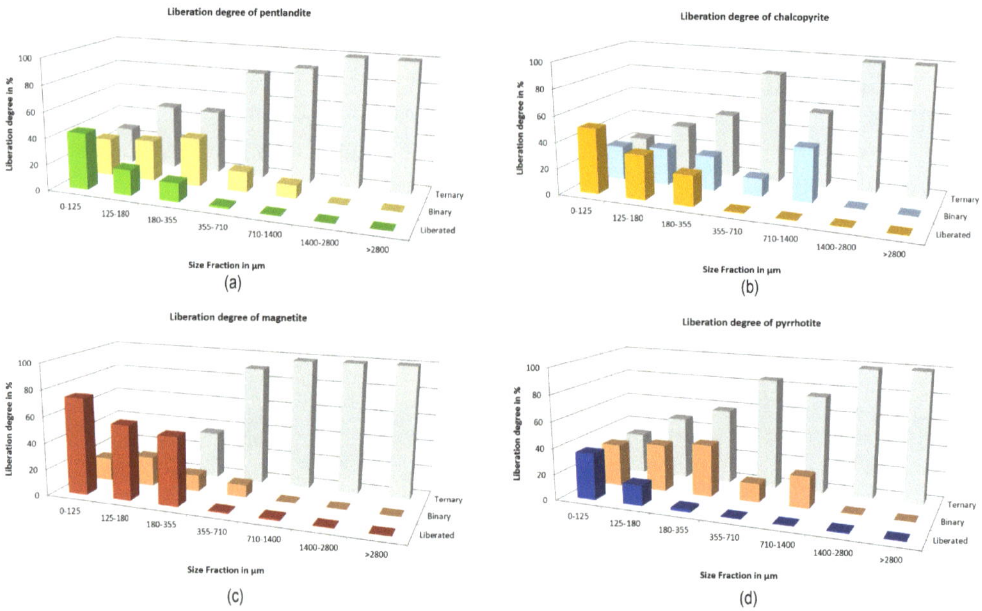

Figure 16. Liberation degree (based on 100% liberation) (**a**) pentlandite, (**b**) chalcopyrite, (**c**) magnetite, and (**d**) pyrrhotite.

In the case of magnetite, the fraction below 125 µm was 73.5% liberated, and just 17.35% was associated with other phases (binary). This increase was visible for fractions below 355 µm in contrast with coarse size classes, larger than 355 µm, where magnetite was associated generally with two phases (Figure 16c). The main phases locked with magnetite were gangue minerals, representing 5.96% in 125/280 μm fractions, which decreased to 2.86% in fractions below 125 μm. Magnetite was also highly locked with pyrrhotite in the fraction below 125 μm (2.36%). For pyrrhotite, the liberation degree was lower than those of valuable minerals, at 35.55% in the fraction smaller than 125 μm (Figure 16d). An important relation in binary particles is the association with pentlandite, which reached 26.15% for the 180/355 μm fraction, decreased to 8.1% for the fraction below 125 μm.

4. Discussion

The evaluation of Ni-Cu samples from Zapolyarnoe in relation to the entire deposit becomes challenging when knowledge of the specific interval of the samples is absent and information on the spatial variation of the deposit is limited. The evaluation of resource estimation and reservoir modelling is not possible, as it is difficult to accurately model the geometry, continuity, and grade distribution of the reservoir without spatial data. Nevertheless, the comprehensive assessment of mineralogical and physical characteristics both before and after comminution has consistently yielded valuable findings for comprehending the material's behaviour and processing.

Upon examining thin sections, significant occurrences of grains with idioblastic textures were observed for rock-forming minerals such as olivine and pyroxene (Figure 6g). In contrast to the exsolution phases (pentlandite, chalcopyrite, and pyrrhotite), these rock-forming minerals show preferential breakage of the mineral aggregates along grain boundaries due to the distinct properties at the boundary contact. The differences include hardness and the sharp boundary contacts, in contrast to the intergrowth between pentlandite and pyrrhotite, for example. After comminution with a jaw crusher, grains containing binary or ternary locking remained in the case of the exsolution phases, mainly pentlandite, chalcopyrite, and pyrrhotite. The locking of exsolution phases after comminution was also evidenced by microindentation at the grain boundaries, where the trend showed that when indenting one of these phases the fracture did not propagate beyond the grain boundary and followed this direction, for example, olivine (Figure 10c).

The consideration of well-studied mineralogical properties is crucial to understanding the mechanical behaviour of minerals. Olivine, feldspar, and pyroxene exhibit remarkably high hardness values, in contrast to pentlandite, chalcopyrite, and pyrrhotite, which exhibit lower hardness values. On the other hand, magnetite shows a unique behaviour attributed to the presence of both primary and secondary radial cracks. These cracks contribute to its ability to separate from other phases and explain its high fracture toughness despite having a high Vickers Hardness Number. Unlike magnetite, chromium-rich minerals such as chromite tend to have lower fracture toughness.

A comparison can be made with pentlandite, as mentioned previously, which presents a higher fracture toughness value (Figure 9). This higher value despite low VHN could be due to the fact that grains of pentlandite have a large number of intrinsic fractures. Pre-existing microstructural defects in pentlandite were linked with an upsurge in the formation of secondary radial cracks observed during Vickers Hardness Testing. In addition, it was observed that chalcopyrite was generally found interfacing with harder minerals and bordering intergrowth. Pentlandite was found interlocking with pyrrhotite in most of the textural observations, explaining why chalcopyrite had a higher liberation degree, in contrast with pyrrhotite, which remained highly locked with pentlandite.

By comparing the degree of liberation with the mineral phase content per fraction (Figure 17a), the increase was accentuated in the fractions smaller than 355 μm, with a decrease between 355 and 710 μm in terms of the content observed for chalcopyrite. When comparing the phase contents of each fraction (Figure 18a–d), they did not vary significantly, except in the case of magnetite.

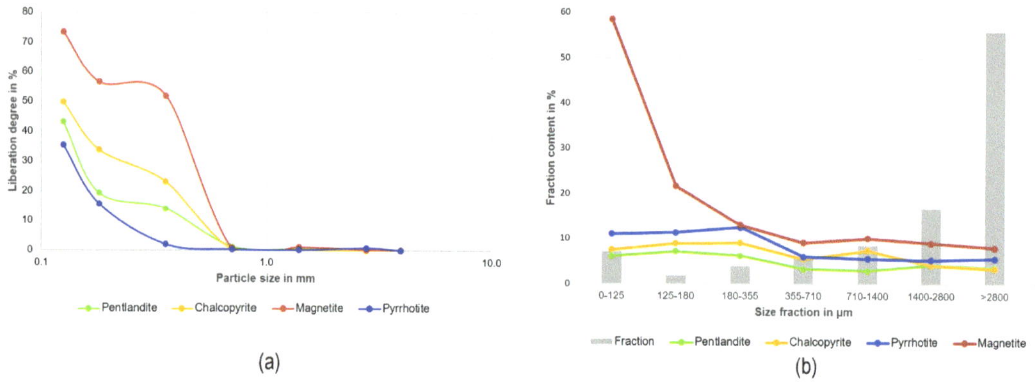

Figure 17. (**a**) Fraction content in percentages in relation to minerals of pentlandite, chalcopyrite, pyrrhotite, and magnetite. (**b**) Degree of liberation of pyrrhotite, pentlandite, chalcopyrite, and magnetite with respect to particle size.

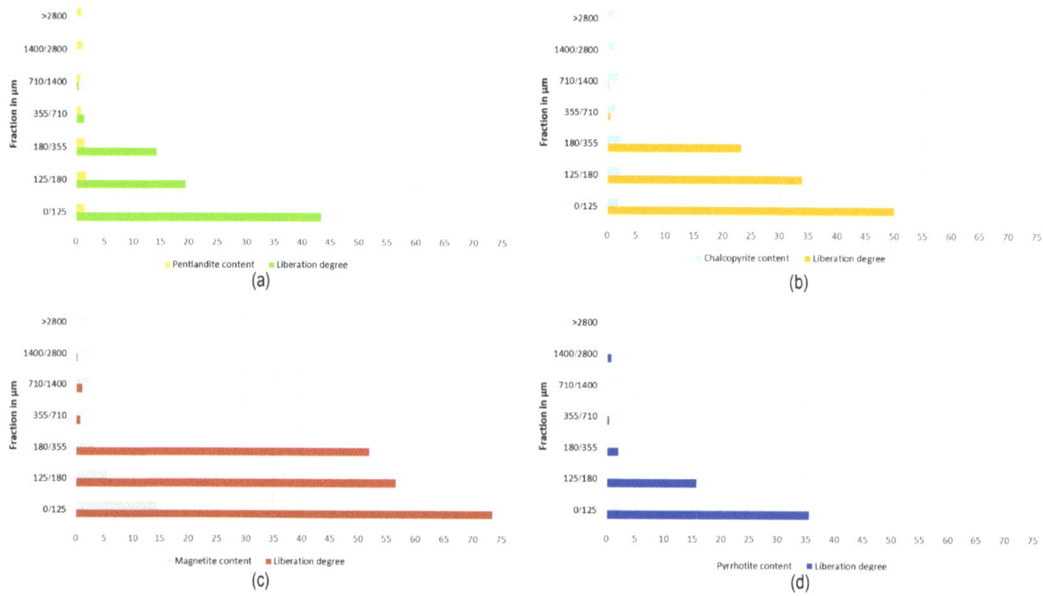

Figure 18. Comparison of liberation degree and mineral content of (**a**) pentlandite, (**b**) chalcopyrite, (**c**) magnetite, and (**d**) pyrrhotite.

When considering the degrees of liberation of pyrrhotite, pentlandite, chalcopyrite, and magnetite with respect to particle size, it can be seen that the chance to separate fully liberated magnetite is high in the small fraction (73.5% liberated) (Figure 17b). A separation possibility, due to the high liberation of magnetite, could be magnetic separation. Strongly magnetic minerals, such as magnetite, Fe-Ti oxides (ilmenite), and iron sulphides (pyrrhotite), can be removed from gangue via low-intensity magnetic separation [54–57]. Examples of magnetic separation tools include discs, drum and roll separators, and overhead magnets.

The investigations carried out on the Zapolyarnoe samples specifically addressed the topics of mineral liberation and preferential breakage in ore mineral processing. The findings revealed that certain rock-forming minerals, such as olivine and pyroxene, possess characteristics that render them suitable for selective comminution at least when using compression stress from jaw crushers. Additionally, magnetite, due to its high fracture toughness, exhibits favourable properties for processing, particularly in coarser fractions (coarser than 355 µm) where full liberation is not achieved. In contrast, exsolution phases including pentlandite, chalcopyrite, and pyrrhotite demonstrate limited suitability for selective comminution. However, in accordance with expectations, the jaw crusher study confirmed increases in the grades of the key mineral phases (pentlandite and pyrrhotite) in the finer fractions, reaching values of 1.8 and 2.3 wt% respectively (Figure 17a).

In this study, a distinction was observed for the prevalent mineral found in copper–nickel deposits. While pyrrhotite is usually the predominant mineral or is found in a relatively similar ratio to magnetite (including Sudbury, Tati Phoenix, and Nkomati) [58–60], magnetite was found to be the most abundant mineral in this particular case.

Despite pyrrhotite exhibiting lower flotation activity compared to chalcopyrite and pentlandite, its high content in the ore poses challenges for selective flotation of minerals [58,59]. When pyrrhotite enters copper and nickel concentrates, it diminishes their quality, leading to significant difficulties in managing the excessive release of sulphur dioxide during the metallurgical process. By separating pyrrhotite into a distinct product, the quality of a nickel concentrate can be enhanced, and the burden on metallurgical processes can be reduced. In theory, it is possible to produce separate concentrates of nickel (pentlandite), copper (chalcopyrite), and iron. However, achieving a clean separation of pentlandite from pyrrhotite is challenging in practice. Utilising ferromagnetic properties or flotation, as described, appears to be the most practical option in this case.

This investigation emphasises the importance of considering the relationship between mineralogical parameters and physical properties when interpreting the enrichment of the ore. Physical observations such as mineral association, grain boundary characteristics, intrinsic fractures, and particle size are crucial to understanding the selectivity potential and product quality because they allow the correlation of textural parameters with expected behaviour and thereby reduce processes that require high energy intensity while seeking the best way to separate the desired phases. A jaw crusher is suitable for crushing material with different strengths quickly. An enrichment at 400 µm can be shown, although it cannot be determined whether the jaw crusher is the most suitable comminution machine to exploit the selectivity potential of the material.

5. Conclusions

The evaluation of Ni-Cu polymetallic ore samples obtained from drilling ore in Zapolyarnoe (Russia) provided the following findings:

- Even from a small amount of drill core material, valuable information for mineral processing can be derived;
- The results of this study demonstrate that certain mineral phases exhibit enrichment, which is related to the physical properties of the rock and grain boundary behaviour. The importance of considering the relationship between mineralogical parameters and physical properties when interpreting ore concentration is emphasised by this study;
- The degree of liberation generally exhibits variation in relation to particle size, where smaller fractions tend to have a greater likelihood of achieving complete liberation of magnetite and, to a lesser extent, chalcopyrite, pentlandite, and pyrrhotite;
- Certain rock-forming minerals, such as olivine and pyroxene, exhibit intrinsic features for preferential breakage, while exsolution phases such as pentlandite, chalcopyrite, and pyrrhotite demonstrate limited suitability;
- Magnetite was found with higher content than pyrrhotite in the copper–nickel ores from the Zapolyarnoe deposit in this case study, which differed from the common predominance of pyrrhotite in other classical Ni-Cu deposits, for example, Sudbury in Canada;

- The variation of pyrrhotite/magnetite poses challenges for flotation due to pyrrhotite's negative impact on concentrate quality and sulphur dioxide emissions during the metallurgical process compared to magnetite. The high presence of chromium in magnetite explains the high value of fracture toughness;
- MLA demonstrated the efficient crushing of minerals with varying hardness values and accurate assessment of enrichment for Cu-Ni fractions below 400 μm. However, it should be noted that a jaw crusher may not be the optimal tool for evaluating selectivity potential due to the relatively small increases in the concentration of the target minerals;
- Further studies and alternative methods, such as utilising ferromagnetic properties or flotation, are recommended to achieve improved separation of pentlandite from pyrrhotite and to target magnetite as a potential mineral to maximise the enrichment after comminution.

Author Contributions: Conceptualisation, A.H.B., O.P. and H.L.; methodology, A.H.B. and O.P.; formal analysis, A.H.B. and O.P.; investigation, A.H.B.; writing—original draft preparation, A.H.B.; writing—review and editing, A.H.B., O.P. and H.L.; visualisation, A.H.B.; supervision, H.L. and O.P.; project administration, H.L.; funding acquisition, H.L. All authors have read and agreed to the published version of the manuscript.

Funding: This research is funded by the Deutsche Forschungsgemeinschaft (DFG, German Research Foundation), Grant No. 429570010.

Data Availability Statement: Data will be made available on request.

Acknowledgments: We thank our project partners at the Institute of Physical Chemistry and the Institute of Mineralogy at the TU Bergakademie Freiberg for their technical support and discussion.

Conflicts of Interest: The authors declare no conflicts of interest.

References

1. Norgate, T.; Jahanshahi, S. Low grade ores—Smelt, leach or concentrate? *Miner. Eng.* **2010**, *23*, 65–73. [CrossRef]
2. McLellan, B.C.; Yamasue, E.; Tezuka, T.; Corder, G.; Golev, A.; Giurco, D. Critical minerals and energy-impacts and limitations of moving to unconventional resources. *Resources* **2016**, *5*, 19. [CrossRef]
3. Aleksandrova, T.; Nikolaeva, N.; Afanasova, A.; Romashev, A.; Kuznetsov, V. Selective disintegration justification based on the mineralogical and technological features of the polymetallic ores. *Minerals* **2021**, *11*, 851. [CrossRef]
4. Nkuna, R.; Ijoma, G.N.; Matambo, T.S.; Chimwani, N. Accessing Metals from Low-Grade Ores and the Environmental Impact Considerations: A Review of the Perspectives of Conventional versus Bioleaching Strategies. *Minerals* **2022**, *12*, 506. [CrossRef]
5. Saramak, D. Challenges in raw material treatment at the mechanical processing stage. *Minerals* **2021**, *11*, 940. [CrossRef]
6. Aleksandrova, T.; Nikolaeva, N.; Afanasova, A.; Romashev, A.; Kuznetsov, V. Justification for criteria for evaluating activation and destruction processes of complex ores. *Minerals* **2023**, *13*, 684. [CrossRef]
7. Tromans, D. Mineral comminution: Energy efficiency considerations. *Miner. Eng.* **2008**, *21*, 613–620. [CrossRef]
8. Ballantyne, G.R.; Powell, M.S. Benchmarking comminution energy consumption for the processing of copper and gold ores. *Miner. Eng.* **2014**, *65*, 109–114. [CrossRef]
9. Powell, M.S.; Morrison, R.D. The future of comminution modelling. *Int. J. Miner. Process.* **2007**, *84*, 228–239. [CrossRef]
10. Jeswiet, J.; Szekeres, A. Energy Consumption in Mining Comminution. *Procedia CIRP* **2016**, *48*, 140–145. [CrossRef]
11. Kukolj, I.; Iravani, A.; Ouchterlony, F. Using Small-scale Blast Tests and Numerical Modelling to Trace the Origin of Fines Generated in Blasting. *BHM Berg-Und Hüttenmänn. Monatshefte* **2018**, *163*, 427–436. [CrossRef] [PubMed]
12. Vizcarra, T.G.; Wightman, E.M.; Johnson, N.W.; Manlapig, E.V. The effect of breakage mechanism on the mineral liberation properties of sulphide ores. *Miner. Eng.* **2010**, *23*, 374–382. [CrossRef]
13. Sandmann, D.; Gutzmer, J. Use of Mineral Liberation Analysis (MLA) in the Characterization of Lithium-Bearing Micas. *J. Miner. Mater. Charact. Eng.* **2013**, *1*, 285–292. [CrossRef]
14. Lieberwirth, H. Securing resources for growth—Challenges and opportunities for comminution in mineral processing. In Proceeding of the Conference in Minerals Engineering—Konferens i Mineralteknik, Luleå, Sweden, 7–8 February 2017.
15. King, R.P. Comminution and liberation of minerals. *Miner. Eng.* **1994**, *7*, 129–140. [CrossRef]
16. King, R.P.; Schneider, C.L. Mineral liberation and the batch comminution equation. *Miner. Eng.* **1998**, *11*, 1143–1160. [CrossRef]
17. Hosten, C.; Cimilli, H. The effects of feed size distribution on confined-bed comminution of quartz and calcite in piston-die press. *Int. J. Miner. Process.* **2009**, *91*, 81–87. [CrossRef]

18. Klichowicz, M.; Lieberwirth, H. Modeling of realistic microstructures as key factor for comminution simulations. In Proceedings of the IMPC 2016—28th International Mineral Processing Congress, Québec City, QC, Canada, 11–15 September 2016.
19. Hesse, M.; Popov, O.; Lieberwirth, H. Increasing efficiency by selective comminution. *Miner. Eng.* **2017**, *103–104*, 112–126. [CrossRef]
20. Weng, X.; Li, H.; Song, S.; Liu, Y. Reducing the entrainment of gangue fines in low grade microcrystalline graphite ore flotation using multi-stage grinding-flotation process. *Minerals* **2016**, *7*, 38. [CrossRef]
21. Lois-Morales, P.; Evans, C.; Weatherley, D. Characterising tensile strength and elastic moduli of altered igneous rocks at comminution scale using the short impact load cell. *Powder Technol.* **2021**, *388*, 343–356. [CrossRef]
22. Lieberwirth, H.; Kühnel, L. Particle size effects on selectivity in confined bed comminution. *Minerals* **2021**, *11*, 342. [CrossRef]
23. Djordjevic, N. Image based modeling of rock fragmentation. *Miner. Eng.* **2013**, *46–47*, 68–75. [CrossRef]
24. Popov, O.; Lieberwirth, H.; Folgner, T. Quantitative charakterisierung der festgesteine zur prognostizierung des gesteinseinflusses auf relevante produkteigenschaften und systemkenngrössen. teil 1: Anwendung der quantitativen gefügeanalyse. *AT Miner. Process* **2014**, *7–8*, 76–88.
25. Evans, C.L.; Wightman, E.M.; Yuan, X. Quantifying mineral grain size distributions for process modelling using X-ray microtomography. *Miner. Eng.* **2015**, *82*, 78–83. [CrossRef]
26. Mariano, R.A.; Evans, C.L. The effect of breakage energies on the mineral liberation properties of ores. *Miner. Eng.* **2018**, *126*, 184–193. [CrossRef]
27. Chen, K.; Yin, W. Investigation of liberation properties and mineral fracture mechanisms of iron ores with different mineral grain sizes at different grinding degrees. *Miner. Process. Extr. Metall. Rev.* **2023**, *40*, 2495–2504. [CrossRef]
28. Tolstykh, N.; Brovchenko, V.; Rad'ko, V.; Shapovalova, M.; Abramova, V.; Garcia, J. Rh, Ir, and Ru Partitioning in the Cu-Poor IPGE Massive Ores, Talnakh Intrusion, Skalisty Mine, Russia. *Minerals* **2022**, *12*, 18. [CrossRef]
29. Tolstykh, N.; Shvedov, G.; Polonyankin, A.; Korolyuk, V. Geochemical features and mineral associations of differentiated rocks of the norilsk 1 intrusion. *Minerals* **2020**, *10*, 688. [CrossRef]
30. Kalinin, A.A.; Kazanov, O.V.; Bezrukov, V.I.; Prokofiev, V.Y. Gold prospects in the western segment of the russian arctic: Regional metallogeny and distribution of mineralization. *Minerals* **2019**, *9*, 137. [CrossRef]
31. Yakubchuk, A.; Nikishin, A. Noril'sk-Talnakh Cu-Ni-PGE deposits: A revised tectonic model. *Miner. Depos.* **2004**, *39*, 125–142. [CrossRef]
32. Lightfoot, P.C. Advances in Ni-Cu-PGE Sulphide Deposit Models and Implications for Exploration Technologies. In Proceedings of the Fifth Decennial International Conference on Ore Deposits and Exploration Technology, Toronto, ON, Canada, 9–12 September 2007; pp. 629–646.
33. Starostin, V.I.; Sorokhtin, O.G. A new interpretation for the origin of the Norilsk type PGE-Cu-Ni sulfide deposits. *Geosci. Front.* **2011**, *2*, 583–591. [CrossRef]
34. Broch, E.; Franklin, J.A. The point-load strength test. *Int. J. Rock Mech. Min. Sci. Geomech. Abstr.* **1972**, *9*, 669–676. [CrossRef]
35. Brook, N. The equivalent core diameter method of size and shape correction in point load testing. *Int. J. Rock Mech. Min. Sci. Geomech. Abstr.* **1985**, *22*, 61–70. [CrossRef]
36. International Society for Rock Mechanics (ISRM); Commission on Testing Method; Working Group on Revision of the Point Load Test Method. Suggested method for determining point load strength. *Int. J. Rock Mech. Min. Sci. Geomech. Abstr.* **1985**, *22*, 51–60. [CrossRef]
37. Duryagina, A.; Talovina, I.; Shtyrlyaeva, A.; Popov, O. Application of Computer X-ray Microtomography for Study of Technological Properties of Rocks. *Key Eng. Mater.* **2018**, *769*, 220–226. [CrossRef]
38. Thuro, K. Bohrbarkeit beim konventionellen Sprengvortrieb. In *Münchner Geologische Hefte Reihe B: Angewandte Geologie*; Publisher: Munich, Germany, 1996; p. 149.
39. Raaz, V. Charakterisierung der Gesteinsfestigkeit mit Hilfe eines modifizierten Punktlastversuches. *Z. Geol. Wiss.* **2002**, *30*, 213–226.
40. Chau, K.T.; Wong, R.H. Uniaxial compressive strength and point load strength of rocks. *Int. J. Rock Mech. Min. Sci. Geomech. Abstr.* **1996**, *33*, 183–188. [CrossRef]
41. Alitalesh, M.; Mollaali, M.; Yazdani, M. Correlation between uniaxial strength and point load index of rocks. *Jpn. Geotech. Soc. Spec. Publ.* **2015**, *2*, 504–507. [CrossRef]
42. *ISO 6507-1*; Metallic Materials—Vickers Hardness Test—Part 1: Test Method. Standard International Organization for Standardization: Geneva, Switzerland, 2005.
43. Bravo, A.H.; Popov, O.; Lieberwirth, H. Mineralogical and micromechanical characterization of slags: Investigated on the example of a MnSiFe slag from electric arc furnace. *Min. Rep. Gluckauf.* **2021**, *157*, 546–559.
44. Tabor, D. The physical meaning of indentation and scratch hardness. *Br. J. Appl. Phys.* **1956**, *7*, 159–166. [CrossRef]
45. Rios, C.T.; Coelho, A.A.; Batista, W.W.; Gonçalves, M.C.; Caram, R. ISE and fracture toughness evaluation by Vickers hardness testing of an Al3Nb-Nb2Al-AlNbNi in situ composite. *J. Alloys Compd.* **2009**, *472*, 65–70. [CrossRef]
46. Blumenauer, H.; Pusch, G. *Technische Bruchmechanik*; Deutscher Verlag für Grundstoffindustrie: Leipzig, Germany, 1993.
47. Evans, A.G.; Charles, E.A. Fracture toughness determinations by indentation. *J. Am. Ceram. Soc.* **1976**, *59*, 371–372. [CrossRef]
48. Tanaka, K. Elastic/plastic indentation hardness and indentation fracture toughness: The inclusion core model. *J. Mater. Sci.* **1987**, *22*, 1501–1508. [CrossRef]

49. Gottielb, P.; Wilkie, G.; Sutherland, D.; Ho-Tun, E.; Suthers, S.; Perera, K.; Jenkins, B.; Spencer, S.; Butcher, A.; Rayner, J. Using quantitative electron microscopy for process mineralogy applications. *JOM* **2000**, *52*, 24–25. [CrossRef]
50. Gu, Y. Automated Scanning Electron Microscope Based Mineral Liberation Analysis. An introduction to JKMRC/FEI Mineral Liberation Analyser. *J. Miner. Mater. Charact. Eng.* **2003**, *2*, 33–41. [CrossRef]
51. Schulz, B.; Merker, G.; Gutzmer, J. Automated SEM mineral liberation analysis (MLA) with generically labelled EDX spectra in the mineral processing of rare earth element ores. *Minerals* **2019**, *9*, 527. [CrossRef]
52. Spencer, S.; Sutherland, D. Stereological Correction of Mineral Liberation Grade Distributions Estimated By Single Sectioning of Particles. *Image Anal. Stereol.* **2000**, *19*, 175–182. [CrossRef]
53. International Society for Rock Mechanics (ISRM); Commission on Standardization of Laboratory and Field Tests. Suggested methods for determining the uniaxial compressive strength and deformability of rock materials. *Int. J. Rock Mech. Min. Sci. Geomech. Abstr.* **1978**, *16*, 138–140.
54. Svoboda, J. *Magnetic Methods for the Treatment of Minerals*; Elsevier: Amsterdam, The Netherlands, 1987; p. 50.
55. Svoboda, J.; Fujita, T. Recent Developments in Magnetic Methods of material separation. *Miner. Eng.* **2003**, *16*, 785–792. [CrossRef]
56. Tripathy, S.K.; Banerjee, P.K.; Suresh, N. Magnetic separation studies on ferruginous chromite fine to enhance Cr:Fe ratio. *Int. J. Miner. Metall. Mater.* **2015**, *22*, 217. [CrossRef]
57. Kumar, D.R.; Srinivas, R.D.; Sita Ram, R.P. Magnetic separation studies for a low grade siliceous iron ore sample. *Int. J. Min. Sci. Technol.* **2013**, *23*, 1–5. [CrossRef]
58. Becker, M.; de Villiers, J.; Bradshaw, D. The flotation of magnetic and non-magnetic pyrrhotite from selected nickel ore deposits. *Miner. Eng.* **2010**, *23*, 1045–1052. [CrossRef]
59. Xu, M.; Wilson, S. Investigation of seasonal metallurgical shift at Inco's Clarabelle mill. *Miner. Eng.* **2000**, *13*, 1207–1218. [CrossRef]
60. Boutroy, E.; Dare, S.A.S.; Beaudoin, G.; Barnes, S.-J.; Lightfoot, P.C. Magnetite composition in Ni-Cu-PGE deposits worldwide: Application to mineral exploration. *J. Geochem. Explor.* **2014**, *145*, 64–81. [CrossRef]

minerals

MDPI

Article

SAG Mill Grinding Media Stress Evaluation—A DEM Approach

Murray Mulenga Bwalya [1], Oliver Shwarzkopf Samukute [1] and Ngonidzashe Chimwani [2,*]

[1] School of Chemical and Metallurgical Engineering, University of the Witwatersrand, Johannesburg 2000, South Africa; mulenga.bwalya@wits.ac.za (M.M.B.); osamkut@gmail.com (O.S.S.)
[2] Department of Mining, Minerals and Geomatics Engineering, University of South Africa (UNISA), Florida Campus, Johannesburg 1710, South Africa
* Correspondence: ngodzazw@gmail.com

Abstract: The volatility of commodity prices has obligated primary metal producers to continuously seek ways of cutting costs in mineral processing units. Improving the wear characteristics and reducing the probability of grinding media fracture can potentially reduce production costs. Characterisation of the impact-loading environment and stress induced into the grinding media in SAG mills aids manufacturers in developing grinding media with superior mechanical properties. Such grinding media development emanates from a firm understanding of the SAG process supported by computer modelling tools and well-established engineering designs. The discrete element method (DEM) is a numerical technique for evaluating collision behaviour in particulate systems. This paper discusses the application of the DEM to estimate survivability and stress, induced into grinding media in a SAG mill.

Keywords: SAG mill; DEM; comminution; grinding media; simulation

1. Introduction

Semi-autogenous (SAG) mills are robustly built heavy-duty machines that can handle run-of-mine (ROM) rocks (ore) to produce product sizes that are fine enough for the separation stage. SAG mills are generally designed with a diameter-to-length ratio of approximately 1.5–3 and are charged with both ore and steel grinding balls of up to 150 mm in diameter. With large diameters typically used in SAG mills, generating high impact and intense shear forces in both cataracting and cascading motion, respectively, the likelihood of mill liner and grinding media damage is enhanced [1].

The versatility of SAG mills in terms of feed size that can be handled and product size that can be produced has propelled these machines to be the technology of choice in many new mineral processing circuits, whilst retrofitting these into existing circuits has also been on the rise. In the wake of that, challenges in terms of control, optimisation and grinding media usage have become prevalent. Among the challenges most operators face with grinding media usage in SAG mills today are premature ball and liner fracture, shape distortion and accelerated wear rates. This is probably because SAG mill environments expose the grinding balls to relatively higher-impact loading, wear and corrosion due to both large mill and ball diameter. To improve the mechanical properties of the grinding media during manufacturing, a firm understanding of the grinding media's mechanical properties and the SAG process are fundamental prerequisites. Grinding media manufacturers develop their products through practical experience, whilst SAG operations are supported by well-established engineering designs and computer-based modelling tools.

Among the tools available to control, optimise and understand load behaviour in SAG mills is the discrete element method (DEM). The DEM is a numerical technique applicable

https://doi.org/10.3390/min15040431

to modelling particle interaction, especially in systems where many particles are involved. This tool can be used to model ball collision behaviour in SAG mills and estimate the stress involved in such collisions. This information is very useful in understanding the toughness and hardness requirements of SAG mill grinding media, which will enable manufacturers to develop grinding media that can withstand the high-impact loading environments in SAG mills.

During impact loading in these mills, the grinding media and the ore are subjected to a suite of stressful events that are responsible for ore fracture but have a cumulative effect on steel media, which may lead to eventual catastrophic failure. For some steel balls, failure may occur after a few repeated cycles of these events, while others seem to display unlimited endurance. The variations observed could be a result of pre-existing imperfections or cracks that act as points of high-stress concentration from which rupture propagates. SAG mills rely on impact energy provided by the cataracting fraction of the charge for most of the ore breakage. This type of charge motion demands the use of grinding balls that have improved fracture toughness to avoid catastrophic failure of grinding balls during operation and thus premature fracture of the balls. The premature fracture is especially enhanced by incorrect design of critical mill operational parameters such as mill speed and lifter face angle, which can project balls to fall above the toe position, resulting in rapid damage of steel balls and lifters [2].

Over the years, significant research has been conducted on SAG mills. The studies include the particle movement laws of a SAG mill, among other things and their influence on speed, lining board, fill level ratio, and steel ball diameter [3], the effect of the steel ball ratio, mill fill ratio and the mill speed ratio on the energy consumption of SAG mill [4], the effect of lifter height/face angle (i.e., lifter wear state) and fill level on charge shape for a given rotational speed [5] and the abrasive wear behaviour of grinding media and mill liners for tumbling mills [6]. Ball-to-rock ratio and steel ball collisions, among other SAG mill internal events, have also been successfully assessed using other techniques such as emission sensing [7,8]. The discrete element method (DEM), which has been widely used to simulate various particulate systems [1,9–12], has also been applied in SAG mill simulations. It was combined with industrial data to track particle movement and energy changes in SAG mills [13]. Despite the excellent work conducted so far to optimise SAG mill operation, little attention has been given to the assessment of the stress induced into the grinding media because of load behaviour in SAG mills. Therefore, to bridge this knowledge gap, the DEM was utilised in this work to investigate the effect of the ball-to-ore ratio on the collision behaviour of balls within the SAG mill. This was achieved by predicting the energy transfer and stress induced in the grinding balls during milling.

2. The Discrete Element Method (DEM)

The DEM is a numerical technique for modelling the collision behaviour of particles in many industrial systems. The strength of the DEM lies in its ability to track the interaction history of each individual particle, even in cases where there are many particles involved. The incorporated particle interactions offer detailed descriptions of the breakage dynamics [14]. The DEM consists of a contact model that is implemented to calculate the forces and energy transfer amongst particles at the point of contact, whilst Newtonian Mechanics is implemented to resolve particle motion from the net forces. The general theory of DEM modelling has been well described elsewhere [9,15,16]. In a SAG mill, the technique has been combined with FEM to predict SAG mill liner stresses to quantify interactions between the solid particles and liners and predict the dynamic loading of these collisions [17]. Later, Cleary et al. [18] combined the DEM and smoothed particle hydrodynamics (SPH) to assess incremental damage of particles and slurry rheology and predict the movement and breakage of coarse particles in SAG mills. The extensive use of these models creates a solid foundation for validating the formula used to determine the operation of SAG mills. In this work, the calculation of the conversion of kinetic energy to elastic energy and the dissipation of energy was based on the linear spring-dashpot contact model. The linear spring-dashpot model is described in detail in previous work by Weerasekara et al. [11] and Owen and Cleary [5]. This model allows the modelling of particles in both the normal and tangential directions, with a frictional slider included for the tangential direction. Particles are allowed to overlap during contact to model elastic deformation, and the extent of overlap is used in conjunction with the contact force law to give instantaneous forces from the knowledge of current positions, orientations, velocities and spins of the particles [19]. The normal force, Fn, is determined by both the spring and dashpot combined in Equation (1) [20]:

$$F_n = -k_n\,\Delta_x + C_n\,v_n \tag{1}$$

where Δ_x is the overlap, k_n is the normal spring constant, C_n is the damping coefficient, and v_n is the normal component of the relative velocity at the contact point. C_n is determined from the specified coefficient of restitution for each combination of colliding materials [21].

$$C_n = \frac{2\ln(\epsilon)\sqrt{k_n m_{eff}}}{\sqrt{\ln\left(\epsilon\right)^2 + \pi^2}} \tag{2}$$

where ϵ is the coefficient of restitution and $m_{eff} = \left(\frac{1}{m_1} + \frac{1}{m_2}\right)^{-1}$ is the effective mass of the particles.

The tangential force (F_t) has an incremental spring based on the integrated tangential displacement and a dashpot for inelastic dissipation. The tangential force at the tangential elastic deformation is combined to give [20]:

$$F_t = \min\left(\mu F_n \sum k_t\,v_t\,\Delta t + C_t\,v_t\right) \tag{3}$$

where μ is the coefficient of friction and v_t is the tangential component of the relative velocity at the contact point [20].

The energy dissipated in collisions is captured in the dashpot, allowing it to be accounted for and recorded for later analysis [11]. An algorithm is used to record the data and store them in arrays that can be extracted for output as required by the user.

3. Materials and Methods

Sag Mill Configuration and Simulation Set-Up

This work shows how the ball-rock composition in the SAG mill affects load behaviour and influences the stress the grinding media is subjected to. The ball sizes investigated were the 100- and 125-mm balls, the data of which were generated from the drop weight experiments designed to match the simulated SAG mill. The SAG mill simulations were based on an industrial SAG mill with specifications presented in Table 1 and the lifter profile shown in Figure 1. The simulations were performed using the Particle Flow Code—Discrete Element Method 3D (PFC-DEM 3D) software, Version 5. The average number of interacting particles in a full-mill simulation was about 160,000. Hence, the amount of computational time required to complete such a simulation would have exceeded 5600 h. To reduce the computational time whilst still retaining important simulation data, a slice of the mill was considered. This significantly reduced the number of interacting particles to an average of 14,500, with the computational time significantly reduced to an average of 504 h. Therefore, for each condition, the simulation was performed once, and any averages were calculated from the two mill revolutions of the simulations. The simulations timed for a mill were approximately 5 days

Table 1. Summary of SAG mill specifications.

Parameter	Value	Units
Inside shell diameter	11.6	m
Length of mill slice	0.5	m
% critical speed	72	Nc
Total filling	24	%
Lifters	Hi-Lo, 650 lifter face	
Number of lifters	36	

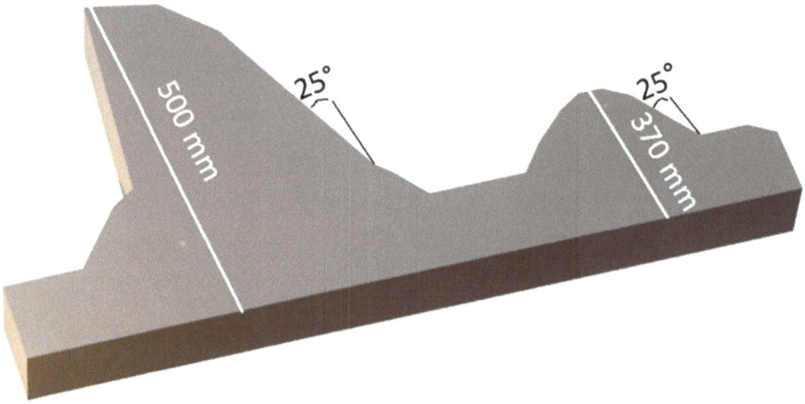

Figure 1. SAG mill lifter profile.

The SAG mill lifters were constructed using moly-chrome steel and have a high-low profile as shown in Figure 1.

The liner configuration in the mill is shown in Figure 2. Table 2 presents different combinations of the ball, ore and total mill filling and ore particle fractions used to simulate the SAG mill. Only the ball-to-ore ratio (by % mill filling degree) was varied, as presented in Table 2, to provide the basis for comparison. Thus, the first simulation (Sim_1) had 100% ball composition, and the ball composition decreased whilst that of the ore filling rose from Sim_2 to Sim_5. The purpose of varying the ball-to-ore ratio was to understand and quantify the impact energy cushioning provided by the ore particles to the grinding balls. Hence, the different SAG mill simulations were compared based on the energy exposed to the grinding balls as a result of the different ball-to-ore ratios, as shown in Table 2.

The simulation was configured to match an industrial mill in terms of internal geometry and operational parameters, but varying the mill speed and ball-to-ore ratio. The ball-to-ore ratio is a very important parameter in operating SAG mills; operating at lower ball loads reduces the frequency of ball-on-ore collisions; hence, breakage rate of the coarse particles will drop, and mill throughput will suffer.

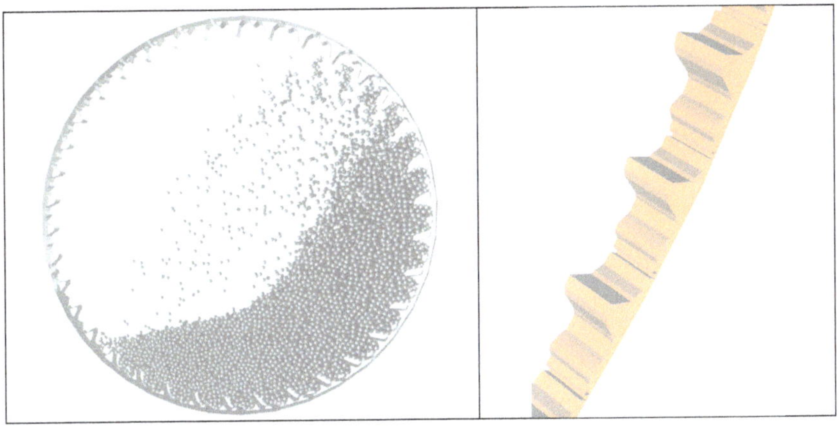

Figure 2. SAG mill simulation liner configuration of end view and 3D view of part of the simulated mill slice.

Table 2. Ball:ore ratios and ore feed distribution in the various mill simulations.

Sim ID	Total Mill Filling (% v/v)	Ball Filling (% v/v)	Ore Filling (% v/v)	Ore Particle Fractions (% v/v)				
				150 mm	120 mm	90 mm	60 mm	30 mm
Sim_1	24	24	0	0	0	0	0	0
Sim_2	24	13	11	2	2	3	2	2
Sim_3	24	10	14	2	3	4	3	2
Sim_4	24	8	16	2	3.5	5	3.5	2
Sim_5	24	6	18	2	2	4	6	4

Higher ball loads are associated with increased mill power draw, overgrinding and an increase in the frequency of steel-on-steel collisions that expose the balls to a high probability of fracture. In this work, the ball-to-ore media ratio was varied to observe the effect of this factor on the stress that the balls are subjected to.

The simulation parameters that were used in this work, presented in Table 3, were adopted from earlier work done by other researchers [15,16,22].

Table 3. SAG mill simulation parameters.

Property	Value	Units
Coefficient of restitution (ϵ)	0.6	
Coefficient of friction (μ)	0.4	
Normal stiffness (kn)	400	kN/m
Shear stiffness (kt)	300	kN/m
Power sampling time interval	0.1	s
Energy scale	5.0	J
Ball size	125	mm

The forces generated in a SAG mill of such geometry are very large and are dependent on the size, mass and tumbling height of the balls.

4. Results and Discussion

4.1. Charge Motion Inside the SAG Mill

During the simulation, various kinds of data were captured at preset intervals for post-simulation review. The data capture had to be optimised to ensure that adequate information was available without bloating the output files. The load behaviour, particle trajectories, mill toe and shoulder positions can be derived from snapshots of all positions captured during simulations and using in-house developed visualisation software, these aspects are easily analysed. The simulation outputs include mill input and dissipated power, energy spectra of all interactive events that involve all particles and other optional information depending on user requirements

The mill and simulation parameters listed in Tables 1 and 3 were kept constant in all simulations except for the mill speed and ball-to-ore ratio, which was varied as described in Table 2. Figure 3 shows the snapshots of the cross-sectional view of the simulated SAG mill. The charge is carried by the lifters along the circumference of the mill in a counterclockwise direction to a point of dynamic equilibrium (shoulder position) where it loses contact with the lifter and either slips along the surface of the cascading stream or falls in free flight until it reaches the bottom of the mill (toe position). For this SAG mill simulation, the shoulder position corresponded to 320° positions as measured anticlockwise from the 12 o'clock position, while the toe was at the 120° position.

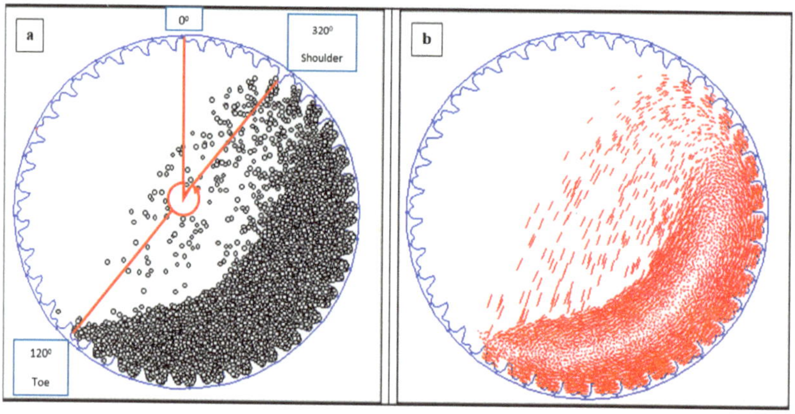

Figure 3. SAG mill load behaviour (**a**) stills (**b**) particle paths.

The end-view snapshots (Figure 3a) show the load behaviour within the mill. Figure 3b is included to accentuate the dynamic behaviour of the load by showing the actual paths followed by the balls.

While the still and particle paths snapshots qualitatively show the load behaviour within the mill, they do not provide information about the type of collisions the particles are involved in (ball-on-ball, ball-on-shell liner or ball-on-ore) and the energy that is associated with such collisions. During the simulation, mill power was recorded as well as the magnitude of each particle contact event, whether a high-energy collision event or a mere touching of the particles. This information was recorded as the energy spectra.

4.2. Mill Power Draw

For mills operating under similar conditions, power variation can be a useful indicator of expected mill performance. It is seen in Figure 4 that the higher the steel ratio, the higher the power draw.

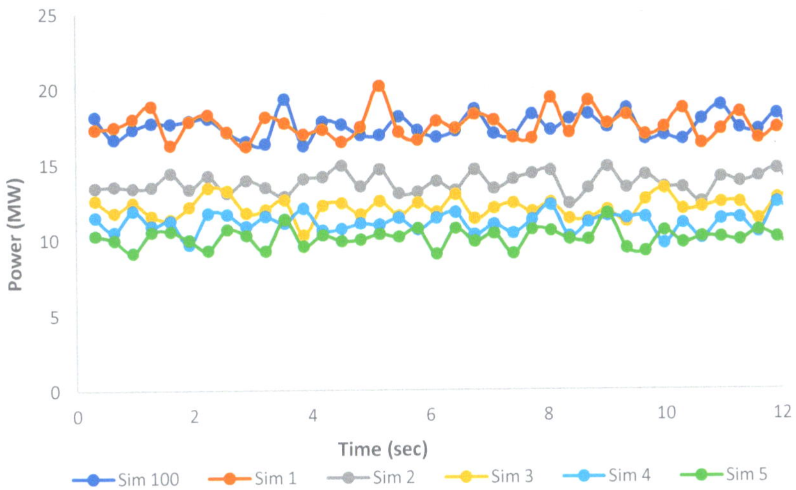

Figure 4. Average mill power draw.

During the simulation, the mill power draw and power dissipation by the balls are recorded. The power is based on the summation of the torque of the balls acting against the mill shell, while dissipated power is based on energy losses by the balls. The dissipated power can be split into damping loss during elastic deformation (represented as overlap in the spring and dashpot model) and frictional loss. Damping loss is the power loss due to both normal and tangential elastic deformation that is modelled by the dashpot in the DEM model. The frictional power loss is due to the rubbing action as sliding occurs between the particle and wall surfaces. The summation of damping and frictional loss gives the total power loss; hence, the load behaviour can be further studied by analysing the split between damping and frictional loss.

Analysis of the power draw split between damping and frictional loss enables the evaluation of the load behaviour within the mill. Power draw dominated by damping loss indicates high-energy trajectories that deliver the much-needed kinetic energy to crush the coarse rocks. This behaviour is mostly associated with SAG mills. Frictional loss domination in the power draw split indicates that the charge behaviour is mostly cascading with low-energy trajectories.

As can be seen in Figure 4, the average power draw is directly proportional to the volume of the mill charge occupied by the steel balls. This is attributed to the difference in density between the rocks (ore) and the steel balls, which makes that of the ore insignificant. In practical mineral processing circuits, the density of steel ranges between 7.75 and 8.00 g/cm^3 whilst ore ranges between 2.40 and 2.85 g/cm^3; hence, the observed power draw trend.

Power draw has been a subject of key research interest, and models have been proposed for mills [23–25] and, more specifically, for SAG mills [26–28]. The Austin model [27], presented in Equation (4), which is reasonably accurate, is used here to compare with the DEM simulated power at different ball-to-ore ratios, and the comparison is presented in Figure 5.

$$P = KD^{2.5}L(1 - AJ_{total})\left[(1-\varepsilon_B)\left(\frac{\rho_{solids}}{w_c}\right)J_{total} + 0.6J_{balls}\left(\rho_{balls} - \frac{\rho_{solids}}{w_c}\right)\right]\phi_c\left(1 - \frac{0.1}{2^{9-10\phi_c}}\right) \tag{4}$$

where A and K are empirical fitting factors, D is the mill diameter inside the liners (m), J_{total} is the mill filling level, as a fraction of total mill volume (e.g., 0.3 for 30%), J_{balls} is the ball fill fraction of the total mill volume, L is the mill effective grinding length (m), P is the power evolved at the mill shell (kW), w_c is the weight fraction of rock in water and rock in the mill, taken as 0.8, ε_B is the porosity of the rock and ball bed, as a fraction of total bed volume, ρ_{solids} is the density of the ore component (t/m^3) and ϕ_C is the mill speed, as a fraction of the mill critical speed (e.g., 0.75 for 75%)

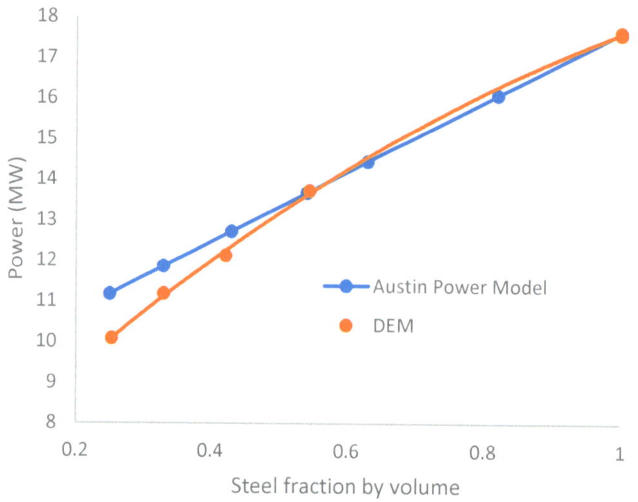

Figure 5. Power trend with increased steel content as predicted by DEM ad Austin SAG power model.

The ore particle and slurry density were kept constant throughout all the simulations. Furthermore, as the fraction of the ore increased in the charge, the observed power draw trend occurred because the small ore particles preferentially sit against the mill lifters, whose ability to lift and throw the steel balls is dependent on the extent of contact between the lifters and the steel balls. As observed by Lameck et al. [29], the position of the toe and the shoulder angles is critical for load behaviour. Ore particles that build up on the lifter surface make the ball roll off the lifters more easily, resulting in an overall lower shoulder position and less cataracting. This results in a lower count of high collision impact as the fine particles in the mill increase.

4.3. Energy Dissipation

Figure 6 shows the trends of the damping and frictional power loss. The summation of the two gives the mill power draw, excluding drive motor inefficiencies and other energy losses such as heat and sound. The mill power draw is dominated by impact loss, which indicates more cataracting occurring than cascading. The cataracting behaviour is characterised by a significant fraction of balls being in free flight after departure from the shoulder. The resulting impacts are high-stress events that break particles at the toe but can equally damage the balls, especially if the cushioning particles are limited.

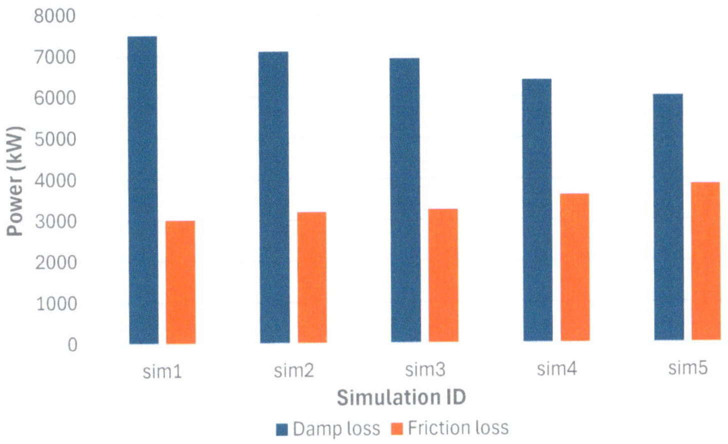

Figure 6. Power draw split between frictional and damping loss as ball fraction increases, excluding motor inefficiencies and other energy losses.

The trend on the frictional energy loss shows an increase in the frictional power loss with an increase of ore (rocks) in the mill. As more rocks are added to the charge, they fracture easily than steel balls, representing a progeny of ore particles varying in size from the largest ore particle fed into the mill to the smallest particle that exits the mill. The small ore particles respond to breakage through abrasion and attrition rather than impact. Hence, the particles fracture and lose weight on account of the rubbing action between ore particles and steel balls, mill liners or other ore particles. This results in more power loss due to friction than damping.

In practical mineral processing grinding mills and engineering systems, an increase in the cataracting fraction of the charge will increase breakage rates of the coarse particles whilst subjecting the grinding balls to significant stress due to the impact loading. On the other hand, if the cascading fraction exceeds the cataracting fraction, the result is more power loss due to friction, finer grinds and increased wear rate of both the grinding balls and mill liners.

While these studies refer to a specific 11.6 diameter industrial mill running at 9 rpm, which is equivalent to 72% of critical speed, the impact of mill speed was explored on the overall mill performance, based on an extreme case of using steel media only. Thus, the effect of speeds 10% lower and 10% higher than the mill under review were assessed. In Figure 7, it is seen that the drive power rises with increasing mill speed. It was also observed that there are fewer power fluctuations for the slower speed of 62% of critical speed. From these results, it is suggested that from a ball preservation perspective, a slower speed would be preferred; however, this would also be dependent on the desirable comminution treatment for the target ore, which is beyond the scope of this work. The impact spectra in Figure 7 also confirm a severe escalation of high-impact energy events with increasing mill speed.

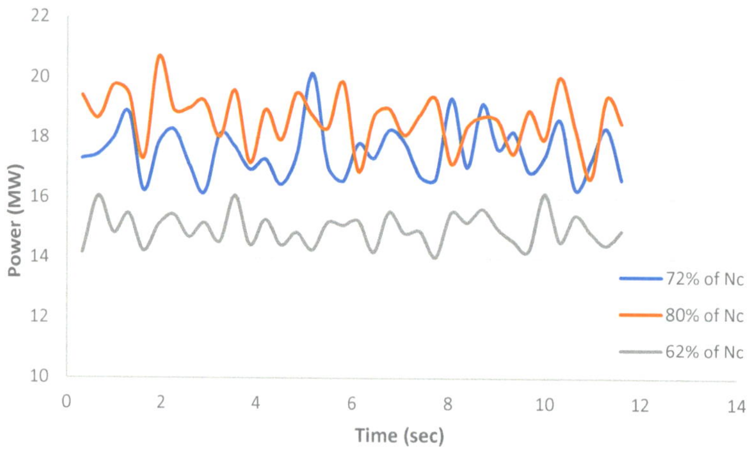

Figure 7. Effect of mill speed on mill drive power.

In Figure 8, the particle position density plots show how the load expands with a cataracting increase as the mill speed increases. For both 62% and 72% of critical speed, the charge is falling within the toe position, thus, there is a high probability of cataracting balls impacting the liner directly; hence, it would not be a recommended option.

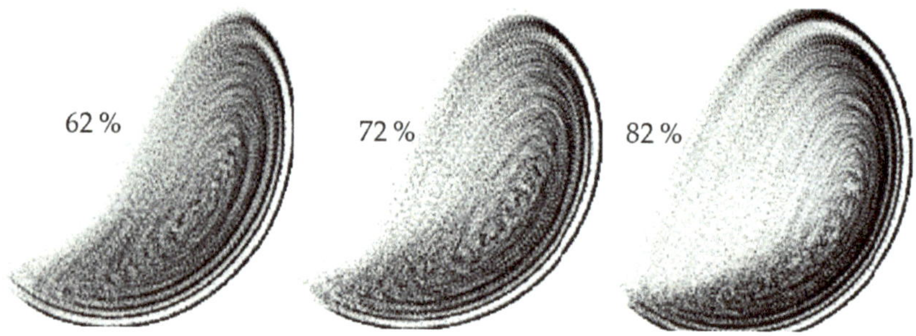

Figure 8. Particle position density plot for mills running at three different mill speeds.

4.4. Impact Energy Spectra (IES)

The IES is a record of the collisions that occurred during the mill simulation period populated into regular energy classes. The size of the energy class is dependent on the geometry of the mill. It should be noted that small energy classes are more definitive than larger classes but at the expense of more computational time. Hence, in this work, a 5 J energy class was selected, which resulted in reasonable computational time whilst the important data, such as the number of collision events and energy involved in collisions, was retained. The technique analyses the prevailing forces within a mill, and its data, combined with the physical ore fracture probability data determined by drop weight tests, can be used to calculate particle breakage within the mill. The use of the IES to determine ore breakage rate has been well described by other researchers [5,30–32]. The IES concept can be extended to evaluate the stress induced into the steel balls resulting from the load behaviour and prevailing forces in the mill. Figure 9 shows the IES of the various simulations conducted for different ball-to-ore ratios.

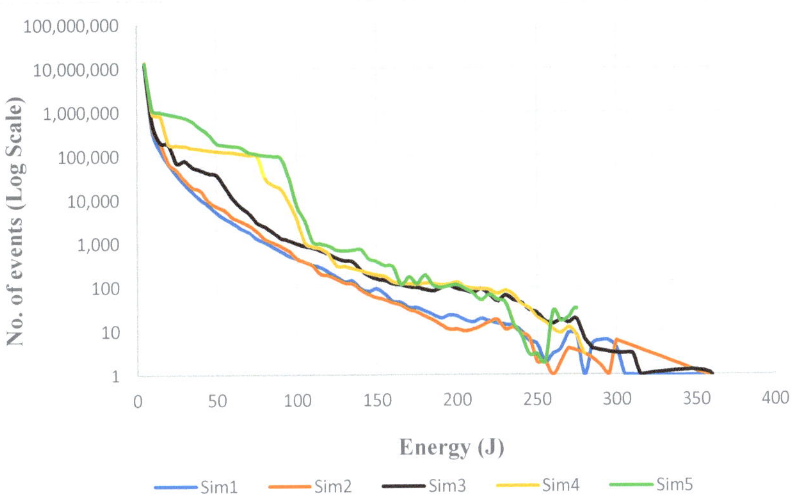

Figure 9. Impact energy spectra of ball-to-ore ratio.

As expected, there are more low-energy than high-energy collision events. The low-energy collisions are mostly common in the cascading region of the mill, whilst the high-energy collisions result from the cataracting fraction of the charge. The maximum collision events occurred in the charge consisting of steel balls only, where the mill shows the highest shoulder position and the trajectories are not cushioned by the ore. This confirms the findings in earlier research [33,34], which showed that the steel ball ratio among the fill level ratio and linear height have a greater influence on the energy distribution in a SAG mill. As the ball-to-ore ratio shifted from mostly steel balls to mostly ore, the number of events in low-energy classes increased while high-energy collision events diminished. This is owing to the difference between the densities of the steel balls and the rocks. Because of the low density of the ore, they tend to cushion some of the impact energy as the balls collide at the toe from free flight. In industrial mineral processing units, the ore cushions the balls from high-energy collisions that can potentially fracture the balls but at the expense of the breakage rate of the large rocks. Furthermore, the build-up of ore particles on the lifter surface results in the earlier media departure from the mill shell. This results in the reduction of both trajectory height and the cataracting fraction of the ore, which thus translates to a high ball wear rate and the inverse proportionality relationship between the impact energy and the ore fraction within the charge.

To further assess the effect of ball diameter on energy, the simulation history of a set of ten (10) 125 mm diameter steel balls was tracked in each of the five simulations described in Table 2. In each simulation, the maximum energy exposed to each 125 mm diameter steel ball was evaluated by averaging the maximum energy that each of the tracked 10 balls was exposed to in that simulation. This was repeated for all the other simulations. Another simulation was performed using only 100 mm diameter steel balls without any ore particles, which replaced the volume of 125 mm balls to determine the effect of ball size on the energy collision spectra. Figure 10 shows the averaged maximum energy that 125 mm diameter and 100 mm diameter steel balls were exposed to in the various simulations.

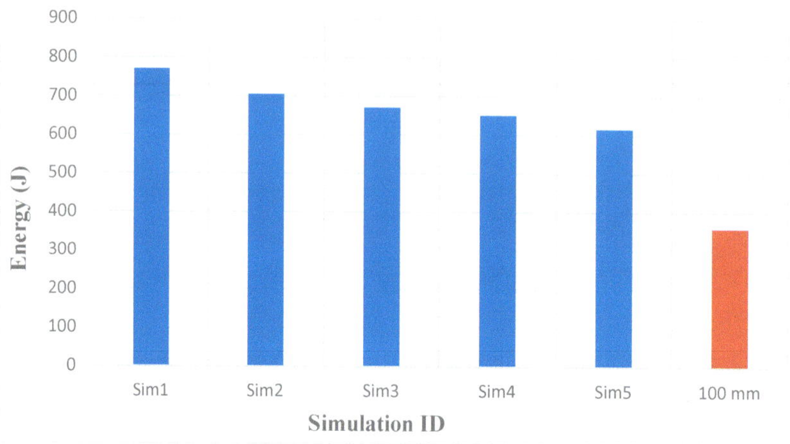

Figure 10. Maximum energy exposed to 125 mm diameter steel balls.

The first simulation shows the highest collision energy events because it was only 125 mm steel balls involved in the collisions without any rocks. As the ball-to-ore ratio decreases, the energy collision events correspondingly decrease due to the cushioning effect provided by the ore, as mentioned earlier. This implies that the use of 125 massive mm balls will have a higher susceptibility to damage, even if cushioning particles are present. The overall effect is that the large size and mass of 125 mm balls exposes them to severe collisions that involve high energy; hence, increasing their probability of fracture more than smaller ball sizes, such as 100 mm balls.

The severity of ball-on-ball and ball-on-wall collisions decreases from 'Sim1' to 'Sim5'. Thus, the optimisation of the SAG mill in terms of the ball-to-ore ratio should aim to achieve the required ore breakage rates at the least possible ball load because:

- lower ball load results in more cushioning of the high-energy collisions that have the potential to fracture the balls;
- a reduction in ball load results in a SAG operation that is more power effective;
- reducing the ball load of the charge will significantly reduce the cost of grinding since grinding media replenishment can account for up to 45% of grinding costs;
- using smaller ball sizes reduces ball damage probability.

Even though the ball-on-ball collisions for the 100 mm balls were aggressive, the highest collision energy recorded was only 387 J. This value is 42% less than Sim5, which constitutes 125 mm diameter steel balls and the highest ore fraction. This clearly shows the significance of ball size on the susceptibility to impact damage. Thus, large balls like the 125 mm diameter steel balls are easily exposed to more severe collisions, even when rock particles are present.

Figure 11 shows the IES of the set of 10 tracked 125 mm diameter steel balls in each simulation. As can be seen in the figure, the tracked balls experience relatively higher energy events with decreased ore fraction. However, occasionally, high-energy events have been observed in higher ore fraction environments, thus making ball breakage possible, albeit less frequently. This is in line with what has been observed in industrial SAG mills with ball filling as low as 3%–5%, where chipped and broken balls were found in the discharge scats. Most SAG grinding operations currently use ball sizes between 100 and 150 mm. However, in most cases, the use of these ball sizes has not been justified by a phenomenological analysis of the real effect on the grinding process but rather by trial and error.

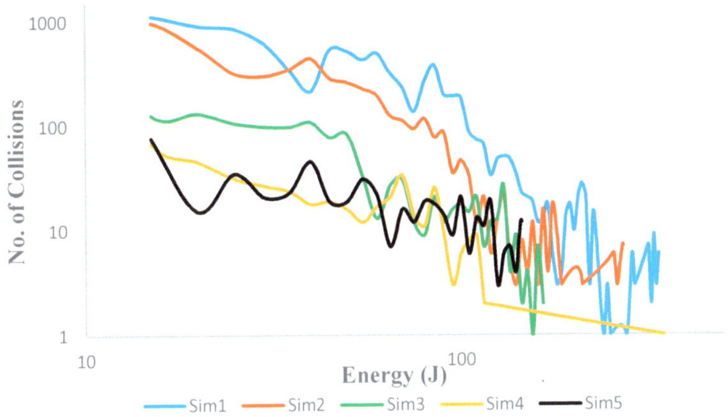

Figure 11. IES of the tracked 125 mm balls.

As for the mill speed, the impact spectra are presented in Figure 12. This is even more definitive as it indicates quantitatively the distribution of stressful impact events. The impact spectra indicate that even the current mill speed is considerably risky if the steel media content were to rise to 100%.

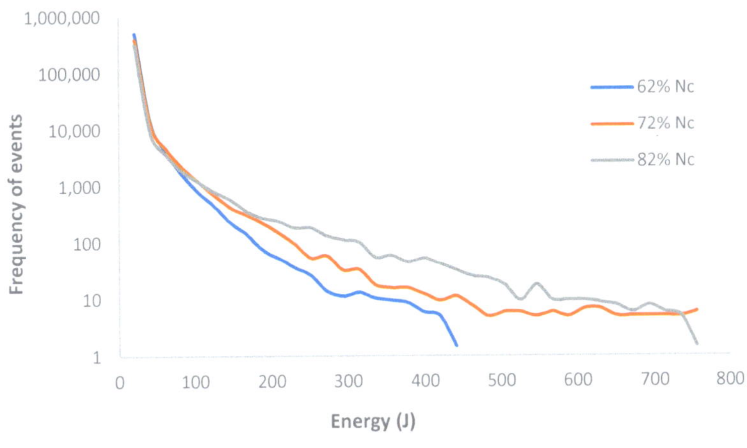

Figure 12. Impact energy spectra of the mill speed.

What is presented in Figures 7, 8 and 12 emphasises that mill speed is an important optimisation criterion for achieving efficient grinding while minimising both ball and liner damage.

4.5. Predicting Fracture of Steel Balls in SAG Mills

Steel ball fracture is not uncommon in SAG mills. The failure of the steel balls is attributed to both the quality of the balls and the energy that they are exposed to. The factors that affect ball mechanical properties include chemical composition, cleanliness of the steel, as well as the type and effectiveness of the heat treatment procedure applied during manufacturing. Usually, the steel ball manufacturers have a firm grip on these factors and ensure that only good-quality balls finally reach the end consumer. However, the environment in which the balls operate, the type and extent of mechanical loading that the balls are exposed to, also have a significant effect on the fracture characteristics of the balls.

Cataracting behaviour exposes both the ore and the grinding balls to impact loading that causes rock fracture. However, the high energy provided by the cataracting load behaviour also exposes the grinding balls to impact loading that stresses the grinding balls until breakage occurs.

To provide some guidelines on how to quantify the frequency of ball damage and breakage expected in a mill environment, reference is made to Equation (5), developed in a study that evaluated the drop weight test [35].

$$Pb = 1 - exp\left[-an^b(X)^c\right] \tag{5}$$

where X is the particle size, n is the number of energy loading events, and a, b and c are model parameters.

Equation (5) can be expressed in terms of energy rather than particle size since the corresponding energy associated with these particles is known and only the values of the parameters change, while the model fit remains the same.

$$Pb = 1 - exp\left[-an^b(E)^c\right] \tag{6}$$

Based on this modified Equation (6), it becomes possible to plot the breakage probabilities as in Figure 13.

Figure 13. Probability of failure based on energy level the steel media is subjected to.

From Figure 13, it can be seen that the higher the cyclic loading energy, the higher the probability of reaching the failure point. For example, if a ball is subject repeatedly to 5000 collision events, there is a 40% chance of failure if the energy level is 800 J, 20% at 400 J and 10% at 200 J. It can thus be appreciated that managing the energy levels is paramount.

From the DBT analysis [35], it was observed that the frequency of collision events increases the probability of ball's eventual failure, but it was clear that pre-existing flaws in the balls are an even more significant factor. Additionally, since the chance of the existence of flaws is higher in big balls, it follows that this will have a compounding effect when large balls are used, as they will have both a high likelihood of flaws as well as get more subjected to higher energy collision events.

The quantification of the stress induced into the steel balls because of impact loading requires knowledge of the energy involved in the cataracting trajectories. The DEM impact energy spectra (IES) thus have proved to be reliable in providing useful information needed to predict the stress induced into the grinding balls.

5. Conclusions

Charge dynamics in SAG mills remain a complex topic for study. However, as this work has demonstrated, the quantification of the stress induced in the steel balls during the SAG process aids grinding media manufacturers in further developing the steel balls so that they can improve their performance capabilities within their service environments. The failure of the steel balls is attributed to both the quality of the balls and the energy to which they are exposed. The environment in which the balls operate, along with the type and extent of mechanical loading they endure, also significantly affects the fracture characteristics of the balls. The analysis of the ball-to-ore ratio and mill speed has shown that larger balls, such as 125 mm balls, have a greater susceptibility to damage, even in the presence of cushioning particles. Moreover, their mass and size make them prone to severe collisions involving high energy, thereby increasing their likelihood of fracture compared to smaller ball sizes, such as 100 mm balls. Both the mill speed and a higher ball-to-ore ratio elevate the shoulder position, which, in turn, enhances the cataracting fraction of the charge. Although this causes the breakage rates of coarse particles to rise, which is advantageous, the grinding balls endure significant stress due to impact loading. In contrast, a decrease in the ball-to-ore ratio results in increased power loss due to friction, low-energy collisions, finer grinds, and a heightened wear rate of both the grinding balls and mill liners. Therefore, it is essential to strike a balance, and that necessitates further research to optimise the charge mixture in SAG mills. Likewise, the information discussed enables SAG mill operators to further understand these machines and the factors that affect the performance, survivability and fracture prediction of the steel balls in such high-impact loading environments. It has also been demonstrated that large balls, such as 125 mm, are prone to fracture even when they comprise a small fraction, and using ball diameters not greater than 100 mm would be recommended to reduce the probability of ball damage in SAG mills.

Author Contributions: Conceptualisation, M.M.B., N.C. and O.S.S.; methodology, M.M.B. and O.S.S.; software, M.M.B.; validation, M.M.B., N.C. and O.S.S.; formal analysis, M.M.B., N.C. and O.S.S.; investigation, O.S.S.; resources, M.M.B.; data curation, M.M.B. and O.S.S.; writing—original draft preparation, O.S.S.; writing—review and editing, M.M.B., N.C. and O.S.S.; visualisation, M.M.B.; supervision, M.M.B. and N.C.; project administration, M.M.B.; funding acquisition, M.M.B. All authors have read and agreed to the published version of the manuscript.

Funding: Scaw metals South Africa.

Data Availability Statement: Data will be shared upon request.

Acknowledgments: The technical assistance and funding support of Scaw Metals South Africa is gratefully acknowledged.

Conflicts of Interest: The authors declare no conflicts of interest.

References

1. Royston, D. Semi-autogenous grinding (SAG) mill liner design and development. *Miner. Metall. Process.* **2007**, *23*, 121–132. [CrossRef]
2. Owusu, K.B.; Zanin, M.; Skinner, W.; Asamoah, R.K. AG/SAG mill acoustic emissions characterisation under different operating conditions. *Miner. Eng.* **2021**, *171*, 107098. [CrossRef]
3. Yang, F.X. Research on Optimization of Working Parameters of SAG Mill. Master's Thesis, Central South University, Changsha, China, 2013.

4. Xie, Q.; Zhong, C.; Liu, D.; Fu, Q.; Wang, X.; Shen, Z. Operation Analysis of a SAG Mill under Different Conditions Based on DEM and Breakage Energy Method. *Energies* **2020**, *13*, 5247. [CrossRef]
5. Owen, P.; Clearly, P.W. The relationship between charge shape characteristics and fill level and lifter height for a SAG mill. *Miner. Eng.* **2015**, *83*, 19–32. [CrossRef]
6. Pourasiabi, H.; Gates, J.D. Ball mill abrasion test (BMAT): Method development and statistical evaluations. *MethodsX* **2022**, *9*, 101900. [CrossRef]
7. Pax, R.A.; Cornish, B. Understanding Size Effects of Semi-autogenous Grinding (SAG) Mill Operation as a Pathway to Solving Feed Disturbances—Case Study Using the MMG Century SAG Mill. In Proceedings of the 13th AusIMM Mill Operators' Conference 2016, Perth, Australia, 10–12 October 2016; The Australasian Institute of Mining and Metallurgy: Perth, WA, Australia, 2016; pp. 321–329.
8. Pax, R.; Thornton, A. Real time measurement and fast control strategies for the optimal operation of grinding circuits. In Proceedings of the MetPlant Conference, Perth, Australia, 10–11 September 2019; pp. 132–147.
9. Sun, Y.I.; Dong, M.; Mao, Y.; Fan, D. Analysis on Grinding media Motion in Ball Mill by Discrete Element Method. In Proceedings of the 1st International Conference on Manufacturing Engineering, Quality and Production Systems, Brasov, Romania, 24–26 September 2009; Volume I, pp. 227–231.
10. Jonsén, P.; Pålsson, B.I. and Häggblad, H.Å.A novel method for full-body modelling of grinding charges in tumbling mills. *Miner. Eng.* **2012**, *33*, 2–12. [CrossRef]
11. Weerasekara, N.S.; Powell, M.S.; Cleary, P.W.; Tavares, L.M.; Evertsson, M.; Morrison, R.D.; Quist, J.; Carvalho, R.M. The contribution of DEM to the science of comminution. *Powder Technol.* **2013**, *248*, 3–4. [CrossRef]
12. Cleary, P.W. and Sinnott, M.D. Simulation of particle flows and breakage in crushers using DEM: Part 1—Compression crushers. *Miner. Eng.* **2015**, *74*, 178–197. [CrossRef]
13. Liu, Y.; Yu, Z.S.; Yang, J.C.; Wassgren, C.; Curtis, J.S.; Guo, Y. Discrete Element Method Investigation of Binary Granular Flows with Different Particle Shapes. *Energies* **2020**, *13*, 1841. [CrossRef]
14. Cleary, P.W.; Sinnott, M.D.; Morrison, R.D. Prediction of slurry transport in SAG mills using SPH fluid flow in a dynamic DEM-based porous media. *Miner. Eng.* **2006**, *19*, 1517–1527. [CrossRef]
15. Mishra, B.K. and Rajamani, R.K. The discrete element method for the simulation of ball mills. *Appl. Math. Model.* **1992**, *16*, 598–604. [CrossRef]
16. Bwalya, M.M. Using the Discrete Element Method to Guide the Modelling of Semi and Fully Autogenous Milling. Ph.D. Thesis, University of the Witwatersrand, Johannesburg, South Africa, 2005.
17. Collinao, E.; Poblete, M.; Tavares, M. Preliminary validation of DEM-FEM coupling to predict mechanical stresses in SAG mill liners. In Proceedings of the International Conference on Semi-Autogenous and High Pressure Grinding Technology, Vancouer, BC, Canada, 20–24 September 2015.
18. Paul, C.; Rob, W.; Morrison, D.; Sinnott, M.D. Prediction of slurry grinding due to media and coarse rock interactions in a 3D pilot SAG mill using a coupled DEM+ SPH model. *Miner. Eng.* **2020**, *159*, 106614.
19. Delaney, G.W.; Cleary, P.W.; Morrison, R.D.; Cummins, S.; Loveday, B. Predicting breakage and the evolution of rock size and shape distributions in Ag and SAG mills using DEM. *Miner. Eng.* **2013**, *50–51*, 132–139. [CrossRef]
20. Cleary, P.W.; Morrison, R.D. Understanding fine ore breakage in a laboratory scale ball mill using DEM. *Miner. Eng.* **2011**, *24*, 352–366. [CrossRef]
21. Govender, N.; Rajamani, R.K.; Kok, S.; Wilke, D.N. Discrete element simulation of mill charge in 3D using the BLAZE-DEM GPU framework. *Miner. Eng.* **2015**, *79*, 152–168. [CrossRef]
22. Polycarpou, A.A. Measurement and Modeling of Normal Contact Stiffness and Contact Damping at the Meso Scale. *J. Vib. Acoust.* **2005**, *127*, 52–60.
23. Bond, F.C. Grinding Ball Size Selection. Mining Engineering. *AIME Trans.* **1958**, *211*, 592.
24. Hogg, R.; Fuerstenau, D.W. Power Relationship for Tumbling Mills. *Trans. Soc. Min. Eng.–AIME* **1972**, *252*, 418–432.
25. Moys, M. A model of mill power as affected by mill speed, load volume, and liner design. *J. S. Afr. Inst. Min. Met.* **1993**, *93*, 135.
26. Loveday, B.K. Prediction of Autogenous Milling from Pilot Plant Tests. In *Proceedings of the 11th Commonwealth Mining and Metallurgical Congress, Hong Kong*; The Institution of Mining and Metallurgy: London, UK, 1978.
27. Austin, L.G. A mill power equation for SAG mills. *Miner. Metall. Process. Soc. Min. Metall. Explor.* **1990**, *7*, 57–63. [CrossRef]
28. Dunne, R.; Morrell, S.; Lane, G.; Valery, W.; Hart, S. Design of the 40 Foot Diameter SAG Mill Installed at the Cadia Gold Copper Mine. In Proceedings of the International Conference on Autogenous and Semiautogenous Grinding Technology 2001, Vancouver, BC, Canada, 30 September–3 October 2001; pp. I-43–I-58.
29. Lameck, N.; Kiangi, K.; Moys, M. Effects of grinding media shapes on load behaviour and mill power in a dry ball mill. *Miner. Eng.* **2006**, *19*, 1357–1361. [CrossRef]
30. Bwalya, M.M.; Moys, M.H.; Hinden, A.L. The use of the discrete element method and fracture mechanics to improve grinding rate prediction. *Miner. Eng.* **2001**, *14*, 565–573. [CrossRef]

31. Rajamani, R.; Callahan, S.; Schreiner, J. DEM Simulation of Mill Charge in 3D VIA GPU Computing. *Miner. Eng.* **2010**, *12*, 1–12.
32. Dabeet, A.; Wijewickreme, D.; Byrne, P. Application of discrete element modeling for simulation of the cyclic direct simple shear response of granular materials. In Proceedings of the 10th National Conference in Earthquake Engineering, Earthquake Engineering Research Institute, Anchorage, AK, USA, 21–25 July 2014.
33. Cleary, P.W. Effect of rock shape representation in DEM on flow and energy utilisation in a pilot SAG mill. *Comput. Part. Mech.* **2019**, *6*, 461–477. [CrossRef]
34. Cleary, P.W.; Owen, P. Effect of operating condition changes on the collisional environment in a SAG mill. *Miner. Eng.* **2019**, *132*, 297–315. [CrossRef]
35. Chimwani, N.; Bwalya, M.; Samukute, O. Using Discrete Element Method to Analyse the Drop Ball Test. *Minerals* **2024**, *14*, 220. [CrossRef]

minerals

MDPI

Article

Calibrating the Digital Twin of a Laboratory Ball Mill for Copper Ore Milling: Integrating Computer Vision and Discrete Element Method and Smoothed Particle Hydrodynamics (DEM-SPH) Simulations

Błażej Doroszuk, Piotr Bortnowski *, Maksymilian Ozdoba and Robert Król

Department of Mining, Faculty of Geoengineering, Mining and Geology, Wroclaw University of Science and Technology, ul. Na Grobli 15, 50-421 Wroclaw, Poland; blazej.doroszuk@pwr.edu.pl (B.D.); maksymilian.ozdoba@pwr.edu.pl (M.O.); robert.krol@pwr.edu.pl (R.K.)
* Correspondence: piotr.bortnowski@pwr.edu.pl

Abstract: This article presents a novel approach to calibrating the digital twin of a laboratory mill used for copper ore milling. By integrating computer vision techniques for real-time data extraction and employing DualSPHysics simulations for various milling scenarios, including balls only, balls with ore, and balls with slurry, we achieve a high degree of accuracy in matching the digital twin's behavior with actual mill operations. The calibration process is detailed for mills with three different diameters, highlighting the adjustments in simulation parameters necessary to account for the absence of ore. Understanding the dynamics between the suspension within the mill and the operation of the grinders is crucial for the future improvement of the grinding process. This knowledge paves the way for optimizing the process, not only in terms of the quality of the end product but primarily in terms of energy efficiency. A profound understanding of these interactions will enable engineers and technologists to design mills and grinding processes in a way that maximizes efficiency while minimizing energy consumption.

Keywords: digital twin; copper ore milling; computer vision; DualSPHysics; DEM simulations; SPH

1. Introduction

The mineral processing industry, a prominent element of modern industrial economies, is crucial for extracting valuable minerals from ores [1]. This fundamental sector transforms raw, naturally occurring materials into useful products, such as metals, which find essential applications in numerous sectors, including construction, automotive, and electronics [2]. In most mineral processing systems, grinding is the one of the first stages, enabling better extraction of valuable components [3], increasing process efficiency, and optimizing energy consumption in subsequent stages of production. In this context, ball mills, integral to mineral extraction and refinement, are central to the fine grinding of ores [4], and the efficacy and efficiency of mineral processing are deeply tied to their performance [5,6].

Despite the simple construction of ball mills and their operation, the grinding process entails a lot of technological challenges. The majority of research efforts and optimization actions are focused on increasing the efficiency of comminution by achieving the proper size reduction in grains [7] with the lowest energy expenditure possible [8]. A particular one is the prediction of mill load behavior, which consists of a mixture of ore particles, water, and grinding media, especially during wet milling. Wet grinding is exceptionally popular in ore processing [9] due to its ability to control the viscosity of the slurry [10], reduce energy consumption and oxidation of certain minerals [11], improve separation [12], and provide control over particle size [13]. The behavior of the mixture, governed by a complex interaction between the liquid, gas, and solid phases inside the mill, is hard to predict [14]. Furthermore, the slurry level in AG/SAG and ball mills significantly influences their power consumption [15].

Contributing to this understanding, ref. [16] conducted a study on the discharge of finer rock, pebbles, ball scraps, and slurry from mills. They demonstrated these elements' effects on overflow ball mills' performance. They combined the Discrete Element Method (DEM) and Smoothed Particle Hydrodynamics (SPH) to model the coarser rock components, grinding media, and slurry components. Their modeling yielded valuable data for understanding and optimizing designs for improved performance and wear life. Similarly, ref. [17] suggested a computational approach that combines the Discrete Element Method (DEM) for modeling the solid particles and a continuum description (CFD) for fluid dynamics in a three-phase model. This approach highlights the usefulness of modeling both charge and slurry dynamics. Moreover, their model considered the effect of drag force on slurry movement evoked by particle motion.

In the field, further advancements include studies on the influence of mill speed, slurry concentration, and slurry filling on the impact forces within the mill and the role of slurry pool formation on mill power draw. Refs. [18,19] used an instrumented ball to measure the charge physical quantities in a ball mill, uncovering complex behavior under different mill speeds and slurry filling levels. In a more recent study, Ref. [20] examined the impact of slurry mass concentration on the grinding characteristics of magnetite using ceramic grinding media, leading to significant energy savings and a reduced chance of over-grinding, especially at high concentrations. Further complementing these insights, Ref. [21] underlined the need to accurately simulate the ore milling process that accounts for the unique shapes of ore particles and grinding media. They proposed a polyhedron-sphere contact model based on the "deepest point method" tested through experiments and simulation studies. They found that their model could lead to increased power consumption, enhanced energy utilization efficiency, and substantial collision energy between the ore and the liner, increasing liner wear.

Moreover, Ref. [22] conducted extensive experimental work to clarify granular flow behavior inside a rotating drum. However, most of these studies involved spherical or nearly spherical particles. The impact of non-spherical particles, such as pharmaceutical tablets suitable as tracer particles, on the velocity profile, residence time, and segregation patterns showed significant deviations from those predicted for spherical particles. In particular, non-spherical particles with an aspect ratio greater than two showed considerable differences, including unexpected core segregation patterns and a lower axial dispersion coefficient.

In other research, Ref. [23] demonstrated the difficulty in modeling multifrequency signals like mechanical vibration and acoustic signals from a wet ball mill during the mineral grinding process. They used the empirical mode decomposition (EMD) technique to decompose these signals into multiscale intrinsic mode functions (IMFs) and developed a new adaptive multiscale spectral features selection approach. Their approach allowed the effective modeling of mill load parameters based on shell vibration and acoustic signals. It provided a practical solution for complex multifrequency signal modeling in mineral grinding.

The presented research works focus on predicting the behavior of various materials undergoing grinding using diverse modeling methods. However, over time, these methods become outdated and require reevaluation to incorporate new components, considering the continuous development of the field. A particular challenge is the lack of a universal approach for modeling the behavior of each slurry, necessitating separate calibration studies based on their composition, with special consideration given to the properties of the ground ore. In this article, building on these insights [2,17,21,23], we aim to develop a numerical model simulating mill load behavior with copper ore slurry. The study utilized laboratory test results obtained from ball mill grinding and subsequently calibrated a numerical mill load model based on these findings. Our ultimate objective is to set a starting point for the further enhancement of our understanding of mill load behavior in both dry and wet milling. We aim to enable more efficient milling strategies, leading to improved performance and reduced energy consumption in the mineral processing industry. The ultimate objective of developing a digital twin is to facilitate numerical testing that predicts the real object's behavior under varied operational parameters, thereby avoiding costly and disruptive changes in an industrial setting. This approach promises significant advancements in operational efficiency and innovation without the need for physical modifications.

2. Theoretical Background

This part provides an overview of the theory surrounding the simulation of the behavior of charge during dry and wet milling with slurry. The necessary elements needed to simulate this behavior include balls, ore, water and steel parts of the mill. In some multiphase approaches, the interaction with gas phase (air) could be also taken into account. A slurry consisting of solid fraction, gas, and water has unique properties that result from the interactions between these three components.

The solid fraction of the slurry inside the mill consists primarily of particles of ground ore, whose size decreases with the progress of grinding. Their size has a significant impact on the slurry's characteristics, as it affects its viscosity and determines the flow behavior [24]. These properties are crucial from the perspective of physics simulation. Additionally, the presence of solid particles can affect the turbidity and transparency of the slurry, but these characteristics are less relevant for the analyzed issue. In addition to the fragmented ore particles, the slurry contains grinding media in the form of steel balls, which complement its composition. Both the balls and ore particles can be modeled as discrete elements with known density, hardness, and size distribution. These parameters have the greatest impact on grinding efficiency and power consumption.

The particle size determines the settling velocity, distribution, conditions of friction between particles, and interactions with other components of the slurry [25]. The density of the ground ore can influence the particles' ability to stay suspended, the resistance to motion, and the distribution of particles in space [26]. The shape of particles determines their velocity distribution and intermolecular interactions, and can affect the dynamics of sedimentation processes under quiescent conditions [27]. The hardness of solid particles refers to their resistance to deformation [28]. This parameter can impact particle–particle interactions and interactions with the mill surface, as well as grinding and comminution processes. The particle size distribution can influence the uniformity of grinding, sedimentation and mixing processes, as well as interactions with other components of the slurry [29].

On the other hand, mill elements can be simulated as one solid body, consisting of all the same properties as discrete elements except for size distribution. The mill lining and grinding media are usually made of steel. The wear and corrosion of these parts can have a significant impact on mill performance.

Water is the main component of the slurry in wet grinding, where it serves as a carrier for solid particles and dissolved gases. Its presence determines the dynamics of particles in the mill. Water can affect the viscosity of the slurry, although its influence is typically smaller compared to that of solid particles [30]. The determining factor for slurry viscosity in this case is its quantity rather than specific physicochemical properties. Additionally, the solubility of gases and the dynamics of certain chemical processes depend on the temperature of the water [31], which is positively correlated with the grinding time due to the release of a significant amount of heat, which is a loss in the grinding process. The main parameters of water that need to be considered in a numerical model are density and viscosity, which depend on its temperature. The density of the slurry affects the balance of forces within it, the behavior of solid particles and gas, and the interactions between them [32]. Viscosity, on the other hand, refers to the flow resistance of the slurry [33]. This parameter is important in modeling the flow of a slurry, as it affects the resistance of solid particles, the friction between slurry components, and the velocity distribution within the system.

2.1. Simulation Environment

DualSPHysics [34] is an advanced open-source code designed to exploit the computational capabilities of modern graphics processing units (GPUs) for simulating real engineering problems using the Smoothed Particle Hydrodynamics (SPH) method. The SPH method, a meshless technique that has significantly progressed over the last two decades, facilitates the simulation of a wide range of physical phenomena. These include the violent hydrodynamics of coastal and offshore structures, galaxy and planetary formation, multiphase flows and mixing in process industries, large deformations of solids and structures, damage and failure modeling, fluid–structure interaction, and applications in computer graphics and games [35–38]. DualSPHysics stands out for its dual capability to operate on both central processing units (CPUs) via the OpenMP shared memory approach and on GPUs, utilizing the CUDA programming framework to achieve computational acceleration.

The development of DualSPHysics is a collaborative effort involving institutions such as the Universidade de Vigo, University of Manchester, University of Lisbon, Università di Parma, Flanders Hydraulics Research, Universitat Politécnica de Catalunya, and New Jersey Institute of Technology. This collaboration has yielded a tool that significantly reduces computational time for SPH simulations, making these simulations accessible on standard desktop PCs. The software, available under the GNU Lesser General Public License (LGPL), includes pre- and post-processing tools and examples and is hosted on GitHub for public collaborative development.

DualSPHysics is specifically optimized for GPU acceleration, featuring CUDA kernels for single GPU use, hierarchical templates, and cell-linked neighbor lists for enhanced flexibility and processing speed [39,40]. The software package also includes dedicated tools for generating inputs and facilitating new test case workflows, along with various post-processing tools for data analysis and visualization.

Since its inception in 2011, DualSPHysics has undergone continuous development, with version 5.0 introducing significant new features such as coupling with the Discrete Element Method [34]. In this research, the newest stable version 5.2.0 was used.

2.2. SPH Formulation

2.2.1. Interpolants and Kernel Functions

In the domain Ω, the continuous integral formulation of Smoothed Particle Hydrodynamics (SPH) for a differentiable function $f(\mathbf{r})$, $\mathbf{r} \in \mathbb{R}^d$, is expressed through the convolution of a kernel function W with f as shown in:

$$\langle f(\mathbf{r})\rangle := \int_{\Omega} f(\mathbf{r}')W(\mathbf{r}-\mathbf{r}',h)d\mathbf{r}'. \tag{1}$$

Here, the $\langle\cdot\rangle$ brackets signify an approximation, $\mathbf{r}' \in \Omega$ serves as an intermediary positional variable, and $W(\mathbf{r},h)$ represents a positive kernel function characterized by the following equation:

$$W(\mathbf{r},h) = \frac{1}{h^d}\omega(q), \tag{2}$$

where $h > 0$ denotes the smoothing length and $q = |\mathbf{r}|/h$ is the normalized distance. The function $\omega : \mathbb{R} \to \mathbb{R}$ is defined as a smooth, non-negative function with properties described by:

$$\int_{\Omega} \omega(q)d\mathbf{r} = \frac{1}{h^d}, \tag{3}$$

This function possesses a compact support as illustrated in: (Figure 1),

$$\omega(q) = 0 \text{ for } |\mathbf{r}| \geq kh, k \in \mathbb{R}^+. \tag{4}$$

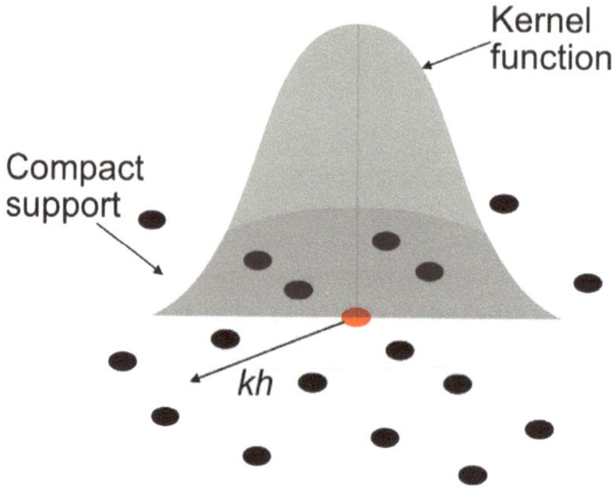

Figure 1. Configuration and compact support of the kernel function (redrawn from [34]).

Numerous smoothing kernels are discussed in the academic context as outlined by [41]. The smoothing kernel is represented by Equation (2) and conforms to the criteria set forth in Equations (3) and (4), with the function $W : \mathbb{R}^d \to \mathbb{R}$. The kernel is characterized by its positive definition over the support domain $\forall \mathbf{r}' \in \Omega; W(\mathbf{r} - \mathbf{r}', h) > 0$, exhibiting a monotonically decreasing and sufficiently smooth behavior over the interval kh, and

$$\lim_{h \to 0} W(\mathbf{r}, h) = \delta(\mathbf{r}), \tag{5}$$

where δ denotes the Dirac delta function as referenced in the works of [42,43]. In the implementation within DualSPHysics, various kernel functions have been utilized. Notably, the third-order B-splines kernel (cubic spline) is prevalent [43]:

$$\omega(q) = a_D \begin{cases} 1 - \frac{3}{2}q^2 + \frac{3}{4}q^3 & 0 \leq q \leq 1 \\ \frac{1}{4}(2-q)^3 & 1 < q \leq 2, \\ 0 & \text{otherwise} \end{cases} \tag{6}$$

where a_D^d is $10/7\pi$ and $1/\pi$ in 2-D and 3-D, respectively.

The fifth-order Wendland function is now popular, characterized by reduced pairing instability [44]. This C^2 kernel exhibits a positive kernel Fourier transform [41,45], and reads

$$\omega(q) = a_d \left(1 - \frac{q}{2}\right)^4 (2q+1) \quad 0 \leq q \leq 2, \tag{7}$$

with a_D^d specified as $10/7\pi$ in two-dimensional spaces and $1/\pi$ in three-dimensional spaces.

The SPH discrete approximation of the convolution between W and f can be described as follows:

$$\langle f(\mathbf{r}_a) \rangle = \sum_{b \in P} f(\mathbf{r}_b) W(\mathbf{r}_a - \mathbf{r}_b, h) \Delta r_b^d, \quad a = 1, \ldots, N, \tag{8}$$

where the subscripts a, b represent the interacting neighboring discrete particles within the set P, which consists of all particles in the domain. This includes particles in the fluid F, boundary B, and any floating objects K as depicted in Figure 2. The set P is the union of F and B, where K is a subset of B. The term ΔV_b denotes the volume V associated with the b-th particle in P. Going forward, the notation $\langle \cdot \rangle$ will be omitted for simplicity.

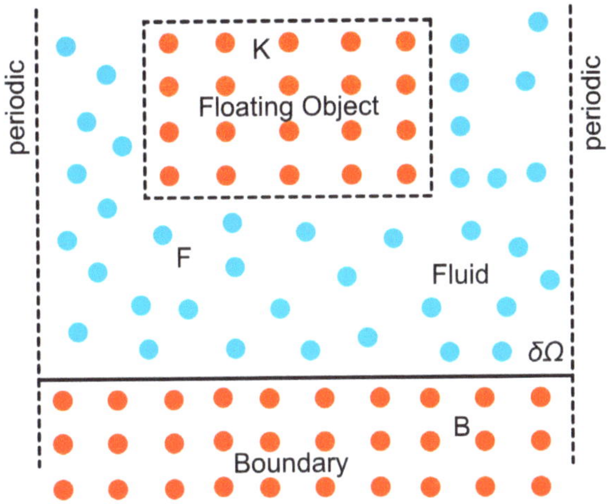

Figure 2. Illustration of various particle sets a ∈ P, with P encompassing all particles within the domain (redrawn from [34]).

For a detailed understanding of SPH, readers are directed to the works of [46,47], along with a comprehensive review found in [48]. Additionally, analogous integral and discrete formulations can be developed for the gradient operator of a specific function f as follows:

$$\frac{\partial f(\mathbf{r})}{\partial \mathbf{r}} = \int_{\Omega} f(\mathbf{r}') \frac{\partial W}{\partial \mathbf{r}}(\mathbf{r} - \mathbf{r}', h)\, d\mathbf{r}', \tag{9}$$

using the convolution theorem,

$$\frac{\partial f(\mathbf{r})}{\partial \mathbf{r}} * W = f * \frac{\partial W}{\partial \mathbf{r}}. \tag{10}$$

The discrete gradient is expressed in the following manner:

$$\frac{\partial}{\partial \mathbf{r}_a} f(\mathbf{r}_a) := \sum_{b \in P} f(\mathbf{r}_b) \frac{\partial}{\partial \mathbf{r}_a} W(\mathbf{r}_a - \mathbf{r}_b, h) V_b, \quad a = 1, \ldots, N. \tag{11}$$

2.2.2. Governing Equations

The fundamental equations governing the behavior of a compressible fluid are encapsulated by the Navier–Stokes equations. These can be expressed in terms of the continuity and momentum equations using a Lagrangian framework as follows:

$$\frac{d\rho}{dt} = -\rho \nabla \cdot \mathbf{v}, \tag{12}$$

and

$$\frac{d\mathbf{v}}{dt} = -\frac{1}{\rho} \nabla P + \mathbf{\Gamma} + \mathbf{f}, \tag{13}$$

where, in this context, d represents the total or material derivative, $\mathbf{v}$ indicates the velocity vector, ρ stands for density, P signifies pressure, $\mathbf{\Gamma}$ refers to the dissipation terms, and $\mathbf{f}$ denotes accelerations caused by external forces like gravity.

2.2.3. SPH Discretization of the Governing Equations

The mass conservation property of SPH,

$$\frac{dm}{dt} = 0, \tag{14}$$

where the mass is conserved exactly within a Lagrangian particle, results in a density change due to the volumetric change of the term at the right-hand side of Equation (12). In the SPH formalism, the discrete form of the continuity equation at point a with position $\mathbf{r}_a$ reads [46],

$$\left.\frac{d\rho}{dt}\right|_{a\in P} = \sum_{b\in P} \frac{m_b}{\rho_b}\mathbf{v}_{ab}\cdot\nabla_a W_{ab} + D_a, \tag{15}$$

where $W_{ab} = W(\mathbf{r}_a - \mathbf{r}_b, h)$ and $(\cdot)_{ab} = (\cdot)_a - (\cdot)_b$. The second term on the right-hand side is indicative of a numerical diffusion term for density [49]. The discrete form of Equation (13) in a SPH formalism reads

$$\left.\frac{d\mathbf{v}}{dt}\right|_{a\in F} = -\sum_{b\in P} m_b\left(\frac{P_a + P_b}{\rho_a\rho_b}\right)\nabla W_{ab} + \langle\Gamma\rangle_a + \mathbf{f}, \tag{16}$$

where a symmetric SPH operator is employed, ensuring the conservation of momentum [50] for the pressure term, and the discrete representation of the dissipation terms is elaborated in the dissipation terms section.

Density Diffusion Terms

Two density diffusion terms (DDT) formulations are incorporated in DualSPHysics, serving as a high-frequency numerical noise filter that enhances the stability of the scheme by smoothing the density and, as a result, the pressure. This improvement is crucial due to the inherent spuriousness in pressure calculations within the weakly compressible SPH framework, which arises from the collocated arrangement (in velocity and density) and explicit approach (in time integration). These terms are described by the general form of,

$$D_a = \delta h_c \sum_{b\in P} \psi_{ab}\cdot\nabla W_{ab} V_b, \tag{17}$$

where δ determines the strength of the diffusion term. The term ψ_{ab} is derived from the Neumann–Richtmeyer artificial dissipation. This form of artificial dissipation was introduced by Molteni and Colagrossi [51] as,

$$\psi_{ab} = 2(\rho_b - \rho_a)\frac{\mathbf{r}_{ab}}{||\mathbf{r}_{ab}||^2}, \tag{18}$$

Recently, Fourtakas et al. [52] modified Equation (18) to include the dynamic component of the density as follows,

$$\psi_{ab} = 2\left(P^{(\mathrm{T})}_{ba} - P^{(\mathrm{H})}_{ab}\right)\frac{\mathbf{r}_{ab}}{||\mathbf{r}_{ab}||^2}, \tag{19}$$

where superscripts (T) and (H) represent the total and hydrostatic components of the density for a weakly compressible and barotropic fluid, respectively, by locally constructing the hydrostatic pressure as,

$$P^{(\mathrm{H})}_{ab} = \rho_0 g z_{ab}. \tag{20}$$

Although both terms exhibit inconsistencies near a truncated kernel support [49] (such as at a free-surface or a wall boundary), resulting in a net outwards vector contribution, the latter term enhances the behavior of pressure near wall boundaries by acting on the dynamic pressure.

Dissipation Terms: Artificial Viscosity

An artificial diffusive term, based on the Neumann–Richtmyer artificial viscosity, can be incorporated into the momentum equation [53] with the objective of diminishing oscillations and stabilizing the SPH scheme. This artificial viscosity term is included within the SPH gradient operator on the right-hand side of Equation (16) and is given by,

$$\Pi_{ab} = \begin{cases} \left(-\alpha \frac{\overline{c_{ab}}}{\overline{\rho_{ab}}}\right) \frac{h\mathbf{v}_{ab}\cdot\mathbf{r}_{ab}}{r_{ab}^2+\eta^2}, & \mathbf{v}_{ab}\cdot\mathbf{r}_{ab} < 0, \\ 0, \quad \mathbf{v}_{ab}\cdot\mathbf{r}_{ab} \geq 0 \end{cases}, \tag{21}$$

where $\overline{(\cdot)_{ab}} = \frac{(\cdot)_a+(\cdot)_b}{2}$. In this context, c represents the numerical speed of sound, and α is the coefficient of artificial viscosity. The parameter $\eta = 0.001h^2$, with $\{\eta \in \mathbb{R}; r_{ab} > \eta\}$, ensures the operator is non-singular. It is demonstrable that $\Pi_{ab} \propto \nu_0 \nabla^2 \mathbf{v}$, where ν_0 denotes the kinematic viscosity [46,54]. The momentum equation, augmented by artificial viscosity, is formulated as,

$$\left.\frac{d\mathbf{v}_a}{dt}\right|_{a\in F} = -\sum_{b\in P} m_b \left(\frac{P_a + P_b}{\rho_a \rho_b} + \Pi_{ab}\right) \nabla W_{ab} + \mathbf{f}, \tag{22}$$

Due to its simplicity, the artificial viscosity formulation is commonly used in SPH as a viscous dissipation term.

Dissipation Terms: Laminar Viscosity

Viscous dissipation of momentum in the laminar flow regime within DualSPHysics is approximated according to Lo and Shao [55],

$$\nu_0 \nabla^2 \mathbf{v}_a = \sum_{b\in P} m_b \left(\frac{4\nu_0 \mathbf{r}_{ab}\cdot\nabla W_{ab}}{(\rho_a+\rho_b)(\mathbf{r}_{ab}^2+\eta^2)}\right)\mathbf{v}_{ab}, \tag{23}$$

where ν_0 represents the fluid's kinematic viscosity. It is noteworthy that the laminar viscous term diverges near the free-surface as elaborated in [56].

Dissipation Terms: Subparticle Scale Model

The large eddy simulation subparticle scale model (SPS) [57] as outlined by Dalrymple and Rogers [58], utilizing Favre averaging within a weakly compressible framework, is incorporated in DualSPHysics. The SPS stress tensor τ is described in Einstein notation across superscripts i, j as per

$$\tau^{ij} = \overline{v^i v^j} - \overline{v}^i \overline{v}^j, \tag{24}$$

modeled by an eddy viscosity closure as,

$$\frac{\tau^{ij}}{\rho} = 2\nu_{\text{SPS}}\left(S^{ij} - \frac{1}{3}S^{ii}\delta^{ij}\right) - \frac{2}{3}C_I\Delta^2\delta_{ij}|S^{ij}|^2. \tag{25}$$

Here, $\nu_{\text{SPS}} = [C_s^2\Delta]^2|S^{ij}|$, where $C_s = 0.12$ is the Smagorinsky constant, $C_\Delta = 0.0066$, Δ represents the initial particle spacing, and $|S^{ij}| = \frac{1}{2}(S^{ij}S^{ij})^{1/2}$, with S^{ij} being an element of the SPS strain tensor. This is utilized alongside the variationally consistent strain tensor.

For the pressure gradient (Equation (16)), the discrete form of the term reads [58],

$$\frac{1}{\rho}\nabla\cdot\tau_a^{ij} = \sum_{b\in P} m_b \left(\frac{\tau_a^{ij}+\tau_b^{ij}}{\rho_a\rho_b}\right)\nabla^i W_{ab}. \tag{26}$$

Finally, the momentum dissipation term in DualSPHysics is expressed as

$$\Gamma_a = \sum_{b\in P} m_b \left(\frac{4\nu_0 \mathbf{r}_{ab}\cdot\nabla W_{ab}}{(\rho_a+\rho_b)(r_{ab}^2+\eta^2)}\right)\mathbf{v}_{ab} + \sum_{b\in P} m_b \left(\frac{\tau_a^{ij}+\tau_b^{ij}}{\rho_a\rho_b}\right)\nabla^i W_{ab}. \tag{27}$$

2.2.4. Equation of State and Compressibility

In the SPH formulation employed in DualSPHysics, density and pressure are interconnected through an equation of state (EOS), which facilitates the weak compressibility of the fluid, predicated on the numerical speed of sound [59],

$$P = \frac{c_s^2 \rho_0}{\gamma}\left[\left(\frac{\rho}{\rho_0}\right)^{\gamma} - 1\right], \tag{28}$$

where γ denotes the polytropic index (typically 7 for water), ρ_0 represents the reference density, and the numerical speed of sound is determined by $c_s = \sqrt{\frac{dP}{d\rho}}$ [60].

A numerical speed of sound c_s is selected based on a typical length scale and timescale of the domain, which permits significantly larger time steps during explicit time integration than what would be feasible with a physical speed of sound [60]. Given $c_s = 10\|\mathbf{v}\|_{\max}$ where $\|\mathbf{v}\|_{\max} = \sqrt{gh_0}$ and h_0 denotes the initial fluid height in the domain, a variation in density of up to about 1% is typically observed. However, this should be closely monitored, as there are exceptional scenarios in which the 1% threshold may be surpassed and compressibility may no longer be considered 'weak'. In such instances, the speed of sound should be increased, leading to a consequent reduction in time step and an escalation in computational time.

2.2.5. Time Integrators and Time Step

In DualSPHysics, two explicit time integration schemes are implemented. For brevity, the governing equations are written as

$$\frac{d\mathbf{v}_a}{dt} = F_a; \quad \frac{d\rho_a}{dt} = R_a; \quad \frac{d\mathbf{r}_a}{dt} = \mathbf{v}_a. \tag{29}$$

The time integrators are briefly introduced below.

Verlet Time Integration Scheme

Verlet schemes [61] are widely adopted in molecular dynamics for their efficiency and the provision of a second-order accurate integration in space. Our approach utilizes a velocity Verlet variant that eliminates the need for multiple computational steps within a single iteration interval. Accordingly, the WCSPH variables are determined as follows:

$$\begin{aligned}
\mathbf{v}_a^{n+1} &= \mathbf{v}_a^{n-1} + 2\Delta t \mathbf{F}_{a'}^n, \\
\mathbf{r}_a^{n+1} &= \mathbf{r}_a^n + \Delta t \mathbf{v}_a^n + \frac{1}{2}\Delta t^2 \mathbf{F}_{a'}^n, \\
\rho_a^{n+1} &= \rho_a^{n-1} + 2\Delta t R_{a'}^n
\end{aligned}$$

Owing to the integration across a staggered time interval, the equations for density and velocity become decoupled, potentially causing the integrated values to diverge. Consequently, an intermediate correction step is necessitated every N_s steps (with $N_s \approx 40$ being advisable) as per the following formulation:

$$\begin{aligned}
\mathbf{v}_a^{n+1} &= \mathbf{v}_a^n + \Delta t \mathbf{F}_{a'}^n, \\
\mathbf{r}_a^{n+1} &= \mathbf{r}_a^n + \Delta t \mathbf{v}_a^n + \frac{1}{2}\Delta t^2 \mathbf{F}_{a'}^n, \\
\rho_a^{n+1} &= \rho_a^n + \Delta t R_{a'}^n
\end{aligned}$$

where the superscript $n \in \mathbb{N}$ denotes the time step and $t = n\Delta t$.

Symplectic Time Integration Scheme

The symplectic position Verlet time integration scheme [62] provides second-order temporal accuracy. It is particularly well suited for Lagrangian methods, owing to its time-reversibility and symmetry when diffusive terms are absent, thus preserving the geometric properties. In the absence of dissipative forces, the position Verlet scheme is given by

$$\mathbf{r}_a^{n+1/2} = \mathbf{r}_a^n + \frac{\Delta t}{2}\mathbf{v}_a^n, \tag{30}$$

$$\mathbf{v}_a^{n+1} = \mathbf{v}_a^n + \Delta t \mathbf{F}_a^{n+1/2}, \tag{31}$$

$$\mathbf{r}_a^{n+1} = \mathbf{r}_a^n + \frac{\Delta t}{2}\mathbf{v}_a^{n+1}. \tag{32}$$

However, with the inclusion of viscous forces and the evolution of density in DualSPHysics, the velocity at the $(n+\frac{1}{2})$ step becomes necessary. Consequently, a half-step velocity Verlet method is employed to calculate the needed velocity for both the acceleration and density evolution at $\mathbf{F}(\mathbf{r}_{n+\frac{1}{2}})$ and $\mathbf{R}(\mathbf{r}_{n+\frac{1}{2}})$, respectively. The approach adopted in DualSPHysics is formulated as

$$\mathbf{r}_a^{n+1/2} = \mathbf{r}_a^n + \frac{\Delta t}{2}\mathbf{v}_a^n, \tag{33}$$

$$\mathbf{v}_a^{n+1/2} = \mathbf{v}_a^n + \frac{\Delta t}{2}\mathbf{F}_a^n, \tag{34}$$

$$\mathbf{v}_a^{n+1} = \mathbf{v}_a^n + \Delta t \mathbf{F}_a^{n+1/2}, \tag{35}$$

$$\mathbf{r}_a^{n+1} = \mathbf{r}_a^n + \Delta t \frac{\mathbf{v_a}^{n+1} + \mathbf{v_a}^n}{2}, \tag{36}$$

where $\mathbf{r}^{n+\frac{1}{2}}$ replaces $\mathbf{r}^{n+1}$ in Equation (30) to remove dependency on $\mathbf{u}^{n+\frac{1}{2}}$. Subsequently, the evolution of density adheres to the half time steps prescribed by the symplectic position Verlet scheme as delineated below [63]:

$$\rho_a^{n+\frac{1}{2}} = \rho_a^n + \frac{\Delta t}{2} R_a^n \tag{37}$$

$$\rho_a^{n+1} = \rho_a^n \frac{2 - \epsilon_a^{n+\frac{1}{2}}}{2 + \epsilon_a^{n+\frac{1}{2}}}, \tag{38}$$

$$\text{where } \epsilon_a^{n+\frac{1}{2}} = -\left(\frac{R_a^{n+\frac{1}{2}}}{\rho_a^{n+\frac{1}{2}}}\right)\Delta t. \tag{39}$$

Variable Time Step

The time integration is constrained by the Courant–Friedrichs–Lewy (CFL) condition, which is essential in explicit time integration schemes to confine the numerical domain within the physical domain of dependence [64],

$$\Delta t_f = \min_a \left(\frac{h}{\left|\frac{d\mathbf{v}_a}{dt}\right|}\right), \quad \Delta t_c = \min_a \left(\frac{h}{c_s + \max_b\left(\frac{|h\mathbf{v}_{ab}\cdot\mathbf{r}_a|}{r_{ab}^2+\eta^2}\right)}\right), \tag{40}$$

$$\Delta t = C_{CFL}\min(\Delta t_f, \Delta t_c), \tag{41}$$

where $\left|\frac{d\mathbf{v}_a}{dt}\right|$ denotes the magnitude of particle acceleration. The variable time step is determined as the minimum between Δt_f and Δt_{cv}, and it is constrained by the Courant number C_{CFL}, typically ranging from 0.1 to 0.2.

2.2.6. Boundary Conditions

Several boundary conditions are implemented in DualSPHysics. Solid boundary conditions are discretized by a set of boundary particles $a \in \mathbf{B}$ that differ from the fluid particles $\mathbf{F}$. Due to the Lagrangian nature of SPH, solid boundaries can also be moved straightforwardly according to user-defined motion.

Dynamic boundary conditions (DBC) are simply represented by fixed particles with density computed with the continuity equation and pressure obtained from the equation of state. Boundaries are easy to set up and computations are relatively stable and efficient, providing a robust option for complex geometries. Validations with dam-break flows and wave flumes have been published, and real engineering applications have been simulated successfully [65,66]. However, an unphysical gap between the fluid and these solid boundaries appears when fluid approaches to the boundary, decreasing the accuracy of pressures measured on the boundary. Therefore, a second approach is included now in DualSPHysics, where the density of solid particles is obtained from ghost positions within the fluid domain by linear extrapolation, described in [67]. With this second approach, the gap is avoided, and pressures of the boundary particles in still water converge to hydrostatic.

Dynamic boundary conditions The dynamic boundary condition (DBC) represents a solid wall using a collection of boundary particles $\mathbf{B}$, where the continuity equation is applied as described in [68]

$$\left.\frac{d\rho}{dt}\right|_{a\in B} = \rho_b \sum_{b\in F} \frac{m_b}{\rho_b} \mathbf{v}_{ab} \cdot \nabla W, \tag{42}$$

while the position of these particles is updated following the equation

$$\left.\frac{d\mathbf{v}}{dt}\right|_{\forall a\in B} = \mathbf{F}^{(\text{imposed})} \tag{43}$$

where $\mathbf{F}^{(\text{imposed})}$ denotes the force exerted on moving boundary particles by the fluid. Such motion can be dictated by a predefined user function or by accelerations accumulated over time. The resultant increase in density from Equation (42) leads to an augmented pressure in the momentum equation for particle $a \in F$, creating a repulsive force between the fluid and boundary particles. This simplified boundary approach is especially suited for GPU implementation, allowing for code optimization through the use of vector lists. Further insights into this method can be found in Crespo et al. [68].

2.2.7. Particle Shifting Algorithm

The anisotropic distribution of particles introduces additional discretisation error through the zeroth- and higher-order kernel moments (i.e., the discrete version of Equation (3)). This is especially true in negligible or large dynamics [59] and violent flows where particles may not maintain an isotropic distribution. In DualSPHysics, the Fickian-based particle shifting algorithm of Lind et al. [69] is used to maintain a near isotropic particle distribution,

$$\left.\delta\mathbf{r}_a\right|_{a\in F} = -D_s \nabla C_a, \tag{44}$$

where $\delta\mathbf{r}_a$ is the shifting distance, ∇C_a is the kernel gradient and D_s

$$D_s = A_s h \|\mathbf{v}_a\| \Delta t, \tag{45}$$

where A_s acts as a control parameter and $\|\mathbf{v}_a\|$ represents the magnitude of velocity for particle a [70]. A constraint based on the magnitude of particle velocity is applied to mitigate the risk of excessive displacement and the subsequent loss of data as particles transition across domain cells. The recommended range for A_s is between one and six, with a preferred value of two as suggested by Skillen et al. [70]. Details for calculating ∇C_a can be found in [34].

2.3. DEM and DCDEM

Interactions between solids and fluids, as well as solid–solid interactions, are crucial issues in various engineering disciplines, including coastal, offshore, maritime, and river engineering. To address these interactions, DualSPHysics incorporates the Distributed Contact Discrete Element Method (DCDEM) into its Smoothed Particle Hydrodynamics (SPH) models for simulating fluid–solid interactions. This method, which calculates forces between a fluid and solid particle pair using SPH and models solid–solid interactions through DEM, is based on an explicit integration method and operates within the meshless framework of DualSPHysics. This innovative approach, initially introduced by Cummins and Cleary, enables the accurate representation of complex solid shapes through a particle-based method, with the particles subjected to a rigid body constraint. With the use of specific non-linear contact models, the system is capable of simulating a wide range of material behaviors using three primary parameters: Poisson ratio, Young's modulus, and dynamic friction coefficient [71–73]. Despite its advantages, the model shares common limitations with general DEM approaches, such as potential stability issues due to its explicit nature, high computational demands or inaccuracies in simulating extended frictional contacts, and the risk of artificial effects in low-resolution simulations caused by the spherical particle shapes. Nevertheless, due to its simplicity and efficiency, it finds applications in evaluating hydrodynamic impacts on structures, assessing natural hazard risks, and designing both floating bodies and exposed structures.

2.3.1. Method Formulation

In the framework of Smoothed Particle Hydrodynamics (SPH), the domain of the fluid is depicted through a constellation of nodal points. These points encapsulate essential physical parameters such as position, velocity, density, and pressure, approximated based on their spatial coordinates. Moving in sync with the fluid, adopting a Lagrangian viewpoint, the properties of these nodes are subject to change over time, influenced by their interaction with nearby nodes. The designation "Smoothed Particle Hydrodynamics" stems from the notion that these nodes essentially mirror the mass of a fluid segment, thereby being termed "particles", and the smoothing over of their motion, ignoring individual rotational dynamics. This methodology is underpinned by the principles of integral interpolation theory [46]. An effective transition from a continuous integral representation to a discrete set of Lagrangian points is achieved through meticulous discretization:

$$A_i \approx \sum_j A_j W(\mathbf{r}_{ij}, h) V_j, \tag{46}$$

known as the summation approximation, is applied across all particles j, for which the distance $|\mathbf{r}_{ij}| = |\mathbf{r}_i - \mathbf{r}_j|$ does not exceed the smoothing length h. Here, V_j represents the volume associated with particle j, A_i denotes the variable being estimated at particle i, and W stands for the kernel, or weight, function. This approach indicates that achieving first-order consistency—wherein the kernel's estimation precisely replicates a linear polynomial function—is not guaranteed. The limitation that arises due to the intrinsic errors linked to the discretized representation

$$\sum_j V_j W(\mathbf{r}_{ij}, h) \approx 1, \tag{47}$$

can become significant, especially near open boundaries or areas of discontinuity where the kernel W fails to maintain compact support. To address this, corrections such as Shepard's method and Moving Least Squares (MLS) are often applied. Following the approach by Colagrossi and Landrini [74], spatial derivatives are determined through the gradient of the kernel. In this study, a Quintic kernel, as proposed by Wendland [44], is utilized:

$$W(\mathbf{r}_{ij}, h) = \alpha_D \left(1 - \frac{q}{2}\right)^4 (2q + 1), \quad 0 \le q \le 2, \tag{48}$$

where $q = |\mathbf{r}_{ij}/h|$ and $\alpha_D = 21/16\pi h^3$ for a 3D case.

2.3.2. Discretization of Rigid Body Equations and Contact Forces with DCDEM

In accordance with the original concept proposed by Koshizuka et al. [75], a rigid entity, denoted as I, is modeled utilizing a group of particles. These particles are arranged such that their relative positions do not alter, effectively allowing the volume of I to be represented by this particle ensemble. For ease of reference, this assembly is also labeled as I. These particles are subject to a unique set of equations distinct from those applied to conventional SPH particles. The dynamics of the rigid body are characterized by Newton's laws, and the process of discretization is achieved by summing up the individual count

$$M_I \frac{d\mathbf{V}_I}{dt} = \sum_{k \in I} m_k \frac{d\mathbf{v}_k}{dt}, \tag{49}$$

$$I_I \frac{d\mathbf{\Omega}_I}{dt} = \sum_{k \in I} m_k (\mathbf{r}_k - \mathbf{R}) \times \frac{d\mathbf{v}_k}{dt}, \tag{50}$$

where the rigid body I is characterized by its mass M_I, velocity $\mathbf{V}_I$, inertia tensor $\mathbf{I}_I$, angular velocity $\mathbf{\Omega}_I$, and center of mass $\mathbf{R}$. These vectorial properties are recalculated at each time step. The force per unit mass acting on a particle k within the particle ensemble that constitutes I is represented by $m_k \frac{d\mathbf{v}_k}{dt}$, where m_k signifies the mass of the particle k. This force integrates various effects including body forces such as gravity, fluid resistance, and the normal or frictional forces experienced during interactions with other solids. As such, the term $m_k \frac{d\mathbf{v}_k}{dt}$ can encompass a range of force types: interactions with fluid particles are governed by the fluid momentum equation, while interactions with solid particles are determined by additional derived forces. This cumulative approach to force calculation facilitates a seamless integration of solid and fluid dynamics.

Utilizing Equation (16) directly for the evaluation of local interactions between fluid and solid phases facilitates the incorporation of a viscous framework, thus providing a mechanism for viscous drag closure [76]. This model operates under the concept of Dynamic Boundary Conditions [68], and it adheres to the specifications outlined in Equations (49) and (50), without the need for any arbitrary, ad hoc adjustments. The dynamics are fully derived from the fundamental, particle-specific solutions of Equations (49) and (50). However, this formulation's challenge lies in the propensity to overestimate density [77,78], a phenomenon linked to an entropy increase at the interface. This leads to an expansion in the spacing between fluid and solid particles due to the exerted pressure gradient force, thereby disrupting the accuracy of viscous force calculations at that interface [74]. The addition of the δ-SPH diffusive term within Equation (15) is suggested as a remedy, enabling an ostensibly precise density estimation at the interface [76].

In the scenario where two particles, either part of a boundary or a rigid body, interact, they exhibit dynamics governed by the Discrete Element Method (DEM). The resultant contact force $\mathbf{F}_i^{\mathrm{T}}$ on particle i, due to a collision with particle j, is analyzed into $\mathbf{F}_n$ and $\mathbf{F}_t$, which correspond to the normal and tangential force components, respectively. These components are further subdivided into a repulsion force $\mathbf{F}^r$, generated by the material's deformation, and a damping force $\mathbf{F}^d$, aimed at energy dissipation throughout the deformation process. The mechanics of DEM interactions between two particles are visually demonstrated in Figure 3.

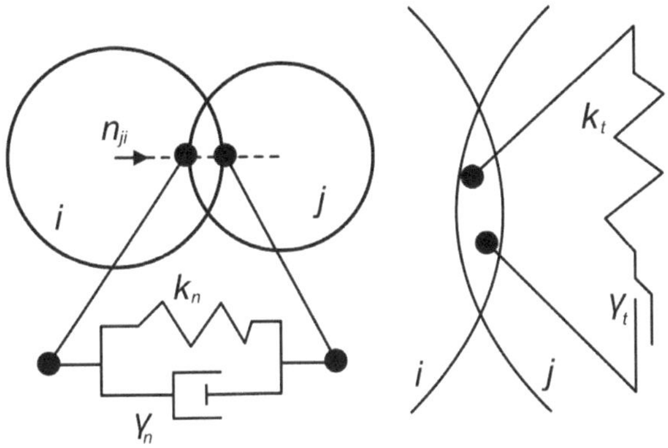

Figure 3. Scheme of DEM mechanism. Left—normal interaction; right—tangential interaction. (Redrawn from [71]).

The normal forces are given by a modified, non-linear, Hertzian model [72,79]

$$\mathbf{F}_{n,ij} = \mathbf{F}_n^r + \mathbf{F}_n^d = k_{n,ij}\delta_{ij}^{3/2}\mathbf{e}_{ij} - \gamma_{n,ij}\delta_{ij}^{1/4}\dot{\delta}_{ij}\mathbf{e}_{ij}, \tag{51}$$

where the stiffness constant for the interacting pair ij is $k_{n,ij}$, and the particle overlap $\delta_{ij} = \max(0,(d_i + d_j)/2 - |\mathbf{r}_{ij}|)$ approximates material deformation. The unit vector between the centers of mass of the particles is denoted as $\mathbf{e}_{ij}$, with the rate of normal deformation given by $\dot{\delta}_{ij} = \mathbf{v}_{ij} \cdot \mathbf{e}_{ij}$, and the damping constant is $\gamma_{n,ij}$. This non-linear formulation helps mitigate some of the limitations inherent in using constant restitution coefficients and fixed contact durations for particle pairs, effectively capturing more physically representative behaviors for a broad spectrum of contacts [80]. The stiffness and damping constants are given by

$$k_{n,ij} = -\frac{4}{3}E^*\sqrt{R^*}; \quad \gamma_{n,ij} = C_n\sqrt{6M^*E^*\sqrt{R^*}}, \tag{52}$$

with C_n of the order of 10^{-5} [81]. The other parameters are given by

$$\frac{1}{E^*} = \frac{1-\nu_I^2}{E_I} + \frac{1-\nu_J^2}{E_J}; \quad R^* = \frac{r_i r_j}{r_i + r_j}; \tag{53}$$

$$M^* = \frac{m_I m_J}{m_I + m_J}, \tag{54}$$

where E is the Young modulus, ν is the Poisson ratio, and m is the mass of the body.

For tangential interactions among particles, friction is depicted through a linear dashpot model, limited by the Coulomb friction law's maximum force, which is defined by the kinetic friction coefficient μ_f. This linear methodology represents a compromise between the fidelity of the model and its computational feasibility. While more elaborate models yield better performance for specific situations, such as rolling friction, their effectiveness is generally limited in broad application scenarios. A sigmoidal function modifies the Coulomb law to ensure continuity at the origin concerning tangential velocity, smoothing the transition and enhancing the model's realism [82]:

$$\mathbf{F}_{t,ij} = \min(\mu_{f,IJ}\mathbf{F}_{n,ij}\tanh(8\dot{\delta}_{ij}^t)\mathbf{e}_{ij}^t \cdot (\mathbf{F}_t^r + \mathbf{F}_t^d), \tag{55}$$

where

$$\mathbf{F}_t^r + \mathbf{F}_t^d = k_{t,ij}\delta_{ij}^t\mathbf{e}_{ij}^t - \gamma_{t,ij}\dot{\delta}_{ij}^t\mathbf{e}_{ij}^t. \tag{56}$$

The model employs a penalty method to replicate the mechanisms of static and dynamic friction. Rather than allowing the body to adhere statically at the point of contact, it is governed by a spring-damper mechanism. With the premise that the time scales for normal and tangential contacts are identical, ref. [83] established that the tangential stiffness could be calculated as $k_t = \frac{2}{7}k_n$. To adequately reflect the time scale of rigid contacts and maintain the system within its stability region, an additional component is required to be incorporated into the dynamic equation as indicated by [81]:

$$\Delta t_{\mathrm{DEM}} = \min_i \left(\frac{\pi}{50} \sqrt{\frac{k_{n,ij}}{m_{IJ}}} \right). \tag{57}$$

This class of DEM model is refereed to as a 'soft sphere' approach since it depends on the particle overlap to deform the spring-damper system. This formulation provides a force estimate particle-wise which, coupled with the rigid body idea, allows the generalization of the geometry by not requiring information on normal directions or other forms of topology, solving for arbitrarily complex geometries by resolving the local interactions.

3. Materials and Methods

This study endeavors to apply the Discrete Element Method (DEM) for the simulation of the grinding media and Smoothed Particle Hydrodynamics (SPH) for the simulation of the slurry. Experiments are performed with the use of copper ore. The experiments consist of 3 stages: calibration of the behavior of the balls; calibration of the dry milling; and calibration of the wet milling. Experiments are recorded with the high-speed camera and then processed with computer vision methods. The simulations are performed with the open-source DualSPHysics simulation frameworks, which consist of a set of C++ and CUDA code, enabling fast simulation using GPU.

3.1. Experimental Setup

The experimental test stand consists of the laboratory scale ball mill with exchangeable drums (Figure 4). In the experiments, 3 different mill drum diameters are used, with external diameters of 300, 400, and 500 mm. The internal diameter of each drum is 8 mm smaller. The laboratory mill is equipped with a transparent wall, allowing for the real-time recording of the milling process. This setup is crucial for the subsequent computer vision analysis. The recordings are made with a speed of 300 fps with a Phantom v2512 camera.

Figure 4. Ball mill with different drum sizes.

3.2. Copper Ore and Its Slurry Properties

The copper ore tested exhibits a density of $2194.93 \pm 98.51\ \mathrm{kg/m^3}$, with the dominant grain size (d_{80}) being approximately 1 mm. This section delves into the critical aspect of viscosity related to the ore and its slurry, presenting it in a manner that aims to be accessible to a diverse audience while retaining the precision expected in scientific discourse.

3.3. Viscosity

Absolute Viscosity: This is measured in poiseuille (Pl), which is equivalent to kg $m^{-1}s^{-1}$ or Newton-second per square meter (N s m^{-2}). It represents the internal friction within a fluid [84].

Kinematic Viscosity: Unlike absolute viscosity, kinematic viscosity does not have a named SI unit and is expressed in square meters per second (m^2s^{-1}). It is obtained by dividing the absolute viscosity by the fluid's density, $\nu = \frac{\mu}{\rho}$, where ν_k is the kinematic viscosity, μ is the absolute viscosity, and ρ is the density of the fluid [84]. It is noteworthy that while the viscosity of gases increases with temperature, the opposite is true for liquids [84].

3.4. Application to Copper Ore Slurry

The slurry used for model calibration presented a density of 1400 kg/m^3. The density was approximated by mixing ore and water based on their known densities, aiming for a reproducible method suitable for industrial scaling, with adjustments made to acknowledge that achieving an exact density is challenging but the target density is closely approximated through direct mixing in specific proportions. Experiments measured the slurry's viscosity across different densities and milling times, utilizing a viscometer AMETEK Brookfield (Figure 5).

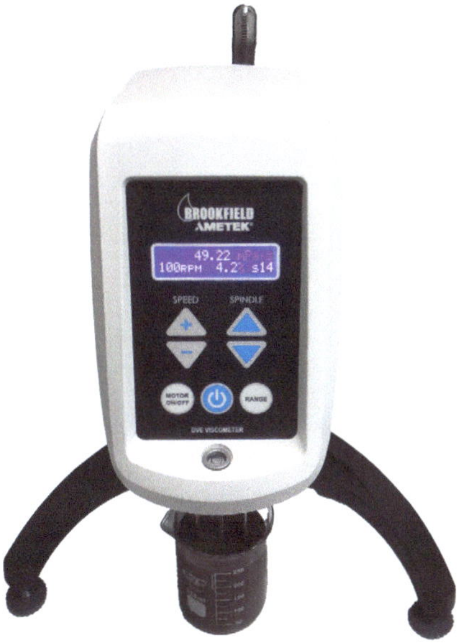

Figure 5. Viscosity measurement of copper ore slurry.

A Random Forest Regressor model was developed to predict the absolute viscosity, ν_a, in Pascal-seconds ($Pa \cdot s$), considering milling time and density as variables. The model, demonstrating an R-squared value of 0.8402, indicates a good fit to the data and reliable prediction capabilities.

For a specific test scenario with a slurry density of 1400 kg/m^3 and a milling time of 15 min, the model predicted an absolute viscosity of approximately 0.01134 $Pa \cdot s$. This provided a realistic prediction of resistance to flow within the slurry.

The kinematic viscosity ν for the copper ore slurry was then calculated by dividing the absolute viscosity by the slurry's density. Given a predicted absolute viscosity of

0.01134 $Pa \cdot s$ and a slurry density of 1400 kg/m^3, the kinematic viscosity of the copper ore slurry was calculated as:

$$\nu = \frac{0.01134}{1400} \approx 8.10 \times 10^{-6}\, m^2/s$$

This positive value of kinematic viscosity reflects the material's resistance to flow and is indicative of the slurry's physical behavior under the given conditions.

3.5. Grinding Media

The grinding media utilized in our experimental setup consisted of steel balls with a makeup diameter of 35 mm. These balls were selected for their uniformity in size and a precisely measured density of 7800.92 ± 55.41 kg/m^3. To ensure a comprehensive understanding of different load behaviors and to enhance the model's adaptability to various grinding media size distributions, we employed a range of ball sizes from 15 mm to 35 mm, with each size differing by 5 mm. The selection of ball sizes was specifically tailored to the drum size utilized in the experiments: for a 500 mm drum, ball sizes ranged from 25 to 35 mm; for a 400 mm drum, from 20 to 35 mm; and for a 300 mm drum, from 15 to 35 mm.

The distribution of steel ball sizes was determined based on the Bond equilibrium state as described by the equation [85,86]:

$$y = \left(\frac{x}{B}\right)^{3.8}$$

where we have the following:

- y represents the percentage of the total equilibrium charge that passes a given size x (mm);
- B denotes the makeup/recharge size of the balls (mm).

3.6. Operational Parameters

The milling operations were conducted with a filling degree of 0.5, wherein 60% of the filled mill's volume was occupied by steel balls, translating to 30% of the mill's total volume given the dynamic nature of milling and the fact that in practice, only 50% of the mill's internal volume is effectively utilized for the balls and the material being milled. The remainder of the volume was considered void space between the balls for calibration purposes in the first step. In dry milling configurations, this void space was filled with ore, while for wet milling, a mixture of water and ore was utilized to achieve the desired slurry density. The selection of the quantity of balls of specific sizes was determined based on the equations detailed in the paper [85]. Figure 6 shows an example of the mill filling in the form of material and grinding media-balls.

Figure 6. Sample ball and ore load.

The rotational velocity of the mill was set to 30% of the critical speed, a value calculated specifically for each experiment to accurately represent the dynamics of the load within the mill drum. It is crucial to note that while the speed was maintained at 30% of the critical speed for all tests, the nominal speed in RPM varied according to the mill diameter: 24.8, 21.1, and 18.7 RPM for diameters of 300, 400, and 500 mm, respectively. This approach ensured that the desired material behavior was achieved without the material being excessively lifted or failing to descend the "slope" created by the grinding media. Achieving consistent behavior across different mill sizes and speed settings, akin to the material's real-world response at 30% of the critical speed, served as an additional layer of validation. The critical speed marks the threshold at which the centrifugal forces acting on the balls equal the gravitational forces, ensuring that the grinding media are carried around in a fixed movement pattern without dropping, a condition essential for efficient grinding.

The critical speed (N_c) in revolutions per minute (RPM) was derived from the balance of gravitational and centrifugal forces, and is given by the formula [87]:

$$N_c = \frac{1}{2\pi}\sqrt{\frac{g}{R-r}} \times 60$$

where we have the following:

- g is the acceleration due to gravity (9.81 m/s^2);
- R is the radius of the mill (in meters);
- r is the radius of the balls (in meters).

This formula is derived from setting the centrifugal force equal to the gravitational force acting on a ball within the mill. This equilibrium point is crucial for determining the optimal operating speed of the mill to ensure the grinding media are the most effective. The value was calculated for the average ball diameter in the load for each experiment.

For all experiments, the mill drums were equipped with eight 5 mm lifters evenly spaced along the interior surface of the drum. This setup was designed to ensure uniform distribution of the grinding media and the material being milled, optimizing the milling efficiency and reproducibility of the results.

The grain composition of the ground ore was standardized, and all experiments were conducted on ore with a grain size below 2 mm, with a d80 grain size of 0.876 mm.

3.7. Computer Vision Analysis

This section details the methodologies applied in analyzing the dynamic angle of repose and the slurry surface position in milling operations through computer vision. The analysis encompasses three scenarios: operations with balls only, dry milling, and wet milling. Each scenario incorporates a blend of image processing techniques and machine learning models to gain insights into the efficiency of the milling process.

Prior to analysis, all scenarios necessitate image pre-processing, which involves loading video, sampling frames at regular intervals. These frames undergo several processing steps using the OpenCV Python library, including mill outline detection via the Hough Circle Transform (Figure 7), centering frames, resizing to match actual dimensions, and applying a mask to obscure everything outside the mill (Figure 8), focusing only on the drum's internal part.

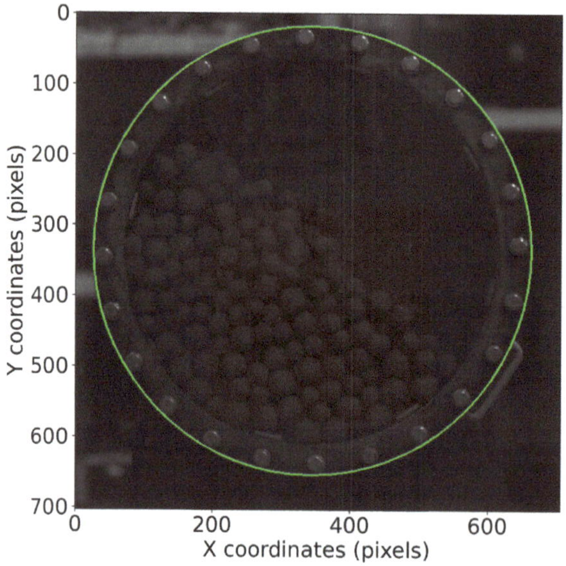

Figure 7. Detection of the mill outline (example mill 500 mm).

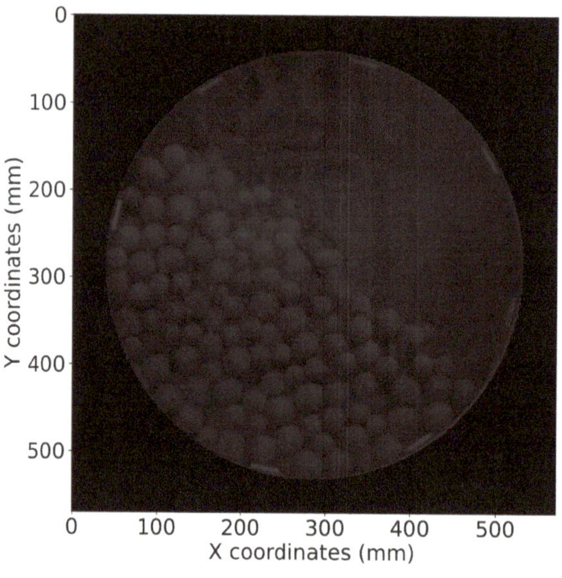

Figure 8. Masked, centered and resized image (example mill 500 mm).

For each recording, a set of 100 frames with equal time step between them is used.

3.7.1. Only Balls

This analysis utilizes the Segment Anything Model (SAM) developed by Meta, a pretrained AI model specializing in segmenting images and applying masks [88]. The methodology unfolds as follows:

1. **Ball Detection and Mask Generation:** Utilize the SamAutomaticMaskGenerator from the SAM library to produce masks for the detected balls (Figure 9).

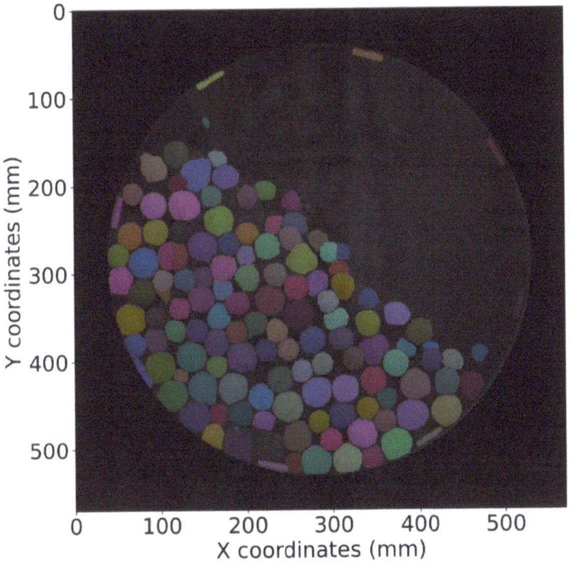

Figure 9. Masks detected with SAM (example mill 500 mm).

2. **Heatmap Generation and Processing:** Generate a heatmap from accumulated binary masks to pinpoint regions with frequent ball presence. Then, to the heatmap, thresholding is applied, and finally the image is smoothed with the smoothing kernel (Figure 10).

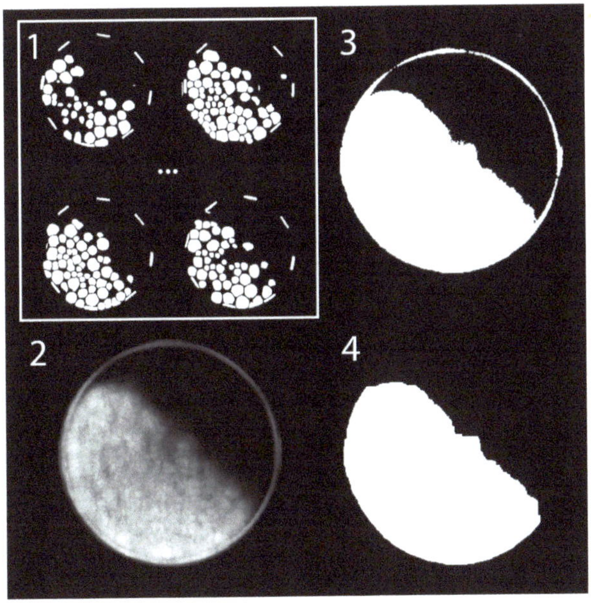

Figure 10. Processing of binary images of mill recording with only balls inside (example mill 500 mm); 1—created set of binary images from detected masks, 2—heatmap created from binary images, 3—heatmap after thresholding, 4—final image after smoothing.

3. **Top Edge Detection and Angle Calculation:** Then processed image is used to detect highest position of the white pixel, and a set of y coordinates for each x coordinate is extracted from the image. Dynamic angle of repose is obtained through linear regression, employing the scikit-learn Python library. Linear regression is performed on the points starting from the 60 mm to the right from the highest detected point, so the most linear part of the repose surface would be analyzed (Figure 11).

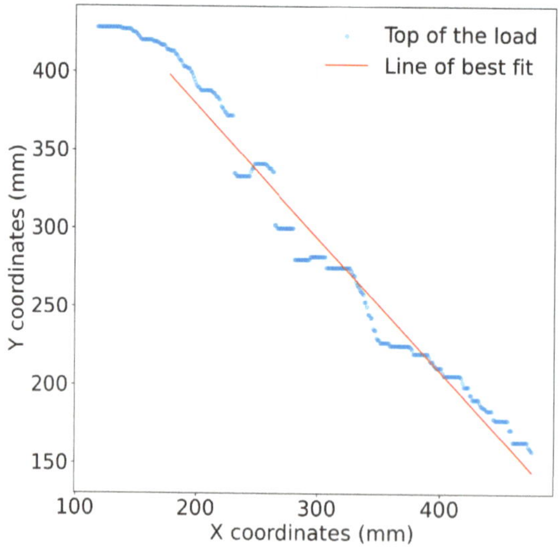

Figure 11. Detected highest position in the Y direction and line fitted with linear regression to the points.

4. **Visualization and Output:** Display the line obtained from the calculated dynamic angle of repose for debugging and threshold adjustment purposes (Figure 12).

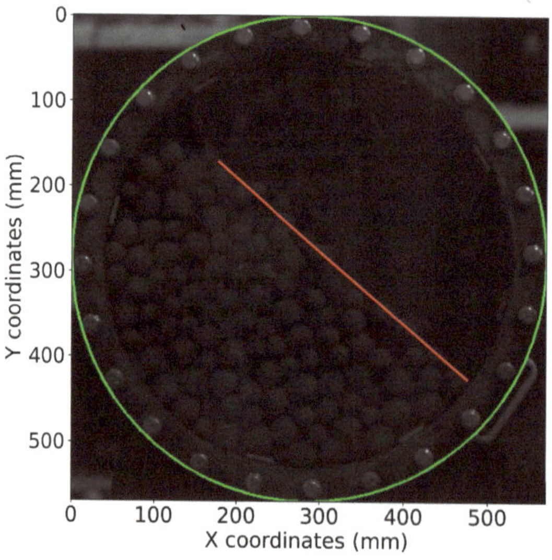

Figure 12. Visualization of detected line over one of the frames from recording (example mill 500 mm).

3.7.2. Dry Milling

This approach differs in the initial detection and isolation of the mill load area and the creation of binary images, reflecting the complexity of processing mixed loads (Figure 13). The difference in this approach is presented in the way that here, thresholding is directly used on the preprocessed image to generate binary images for further analysis.

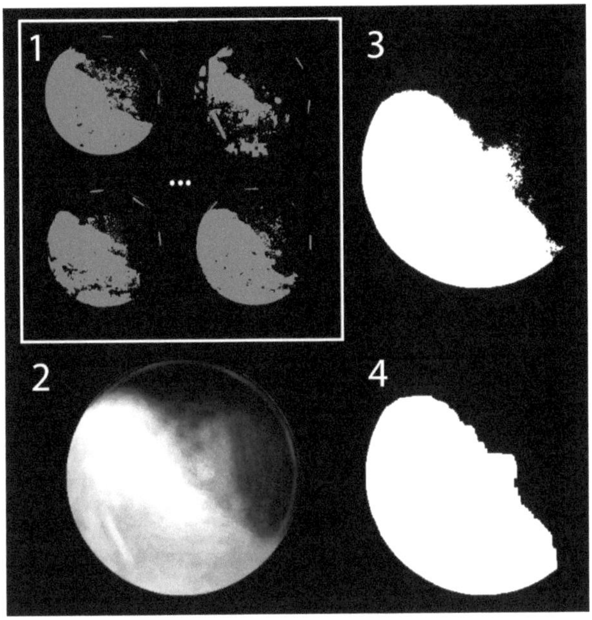

Figure 13. Processing of binary images of mill recording of dry milling (example mill 500 mm); 1—created set of binary images from detected masks, 2—heatmap created from binary images, 3—heatmap after thresholding, 4—final image after smoothing.

3.7.3. Wet Milling

Wet milling operations present unique challenges, making computer vision models unsuitable for extracting the mill load area. Consequently, an interactive GUI interface was developed to address these obstacles. The load, barely visible through the slurry-covered transparent wall, necessitates a procedure combining manual point selection with curve fitting techniques. In this scenario, the top surface of the liquid is analyzed rather than measuring the angle of repose.

Because of the obstacles described before, only the recording of the mill with the biggest drum was suitable for the calibration of the wet milling. The medium and small size drums did not provided enough valuable data to process them the same way.

During milling experiments, the observation was made that when ball load is under the slurry, the whole transparent wall area is equally covered with the slurry and nothing is visible, but when balls appear over the slurry surface, they start to wipe off parts of the slurry from the transparent wall. This phenomenon was manually caught to measure the approximate location of the slurry surface.

1. **Manual Point Selection:** Conduct a manual analysis using a custom interface to select points delineating the wet load's boundary (Figure 14).

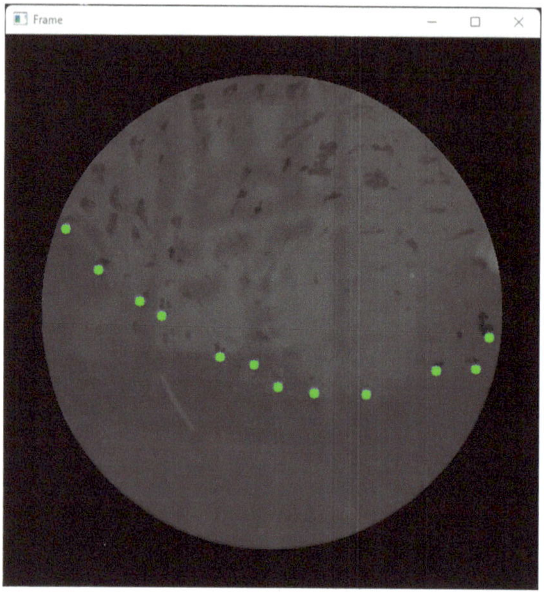

Figure 14. GUI interface (example mill 500 mm).

2. **Data Preparation and Correction:** Prepare and correct the selected points for the analysis coordinate system, emphasizing accuracy verification.
3. **Curve Fitting and Analysis:** Perform curve fitting to model the load's boundary, calculate the fit's parameters, and evaluate the fit's uncertainty.
4. **Visualization and Interpretation:** Display the fitted curve alongside the original data points and confidence interval bounds.

This methodology distinctively blends manual interaction and statistical modeling to analyze wet milling dynamics, utilizing tools such as OpenCV, matplotlib, pandas, and scipy for thorough data management and analysis.

3.8. DualSPHysics Simulation

Simulation in DualSPHysics is set up by defining XML file. Files contain a few important parts, such as the constants definition, parameter definition, geometry definition, movement definition, and definition of the material parameters. Constants and parameters needed to run the simulation are shown below (Table 1 shows the material parameters, and Table 2 shows the simulation settings).

Parameters were selected by varied trial-and-error testing to achieve the stability of the simulation for this specific use case and scale. Material properties for each "bound" (boundary) and "float" (free moving body) were set as steel with a Poisson ratio of 0.29 and Young modulus of $1 \cdot 10^8$ Pa. The Young modulus was reduced to achieve more efficient simulation times. The decrease in the Young modulus was balanced by further calibration. Due to DEM calculations in the simulation, a fixed time step was used because it was observed to provide the best stability and fewest errors during calculations.

Table 1. Material parameters and constants in simulation.

Constant	Value	Description
Lattice	1 (bound), 2 (fluid)	Type of lattice to create initial particles
Gravity	z = −9.81	Gravitational acceleration, m/s^2
Reference Density (ρ_0)	1400.0	Reference density of the fluid, kg/m^3
Density Gradient	Rho0	Initial density gradient method
HSWL	0.8 (dry), Auto (wet)	Max still water level for sound speed calculation.
γ	7 (default)	Polytropic constant for water
System Speed	Auto	Max system speed for calculations
Coef. of Sound	20 (default)	Coefficient to multiply system speed
Sound Speed	Auto	Speed of sound in the simulation
Coef. h	0.15 (dry) 0.8 (wet)	Coefficient for smoothing length calculation
CFL Number	0.2	Coefficient to multiply dt

Table 2. Parameters and settings of the simulation process.

Parameter	Value	Description
SavePosDouble	0 (default)	Precision for saving particle position
Boundary	DBC	Boundary method used
Step Algorithm	Verlet	Step algorithm for integration
Verlet Steps	40 (default)	Steps for Eulerian equations in Verlet
Kernel	Wendland	Interaction kernel type
Visco Treatment	Artificial	Viscosity formulation
Viscosity (ν)	$8.1 \cdot 10^{-6}$	Viscosity value m^2/s
Visco Bound Factor	0.1	Multiplier for boundary viscosity
Density Diffusion Term	Fourtakas	Method for density diffusion
DDT Value	0.1 (default)	Value for density diffusion term
Shifting	Full	Shifting mode for corrections
Shift Coef	−2 (default)	Coefficient for shifting computation
Shift TFS	2.75 (default)	Threshold to detect free surface
Rigid Algorithm	DEM	Algorithm for rigid body dynamics
DtFixed	$1 \cdot 10^{-5}$ s	Fixed time step value, s
DtAllParticles	1	Inclusion of particles for DT calc.
TimeOut	0.01 s	Time out for data output, s
PartsOutMax	1%	Max fluid particles out of domain
RhopOutMin	700 kg/m^3	Minimum valid density
RhopOutMax	2000 kg/m^3	Maximum valid density

3.8.1. Geometry Definition

Geometry in the simulation environment has to be defined also in the XML script. Simplified geometry was used to represent the mill and lifters. The mill was defined as a cylinder, and lifters as boxes with proper dimensions. The resolution of discrete particles was defined as 0.005 m. It is the distance between discrete particles, representing all bodies and liquids in the simulation. To resolve the problem of sphere and liquid packing in the limited space in the initial state of the simulation, the mill was modeled as a long cylinder (Figure 15).

Script in Python was developed to generate the particles and liquid in such a way that they would not overlap anything else and not cause errors when populating the geometry with discrete particles.

One of the mill walls was defined as a moving part that, during the first 2 s of the simulation, pushed all of the load into the final volume of the mill, representing the real volume of the laboratory mill used in experiments. Those first 2 s of the simulation were not taken into account during any analysis. Each simulation had 12 s (2 s for wall movement and 10 s of normal mill work).

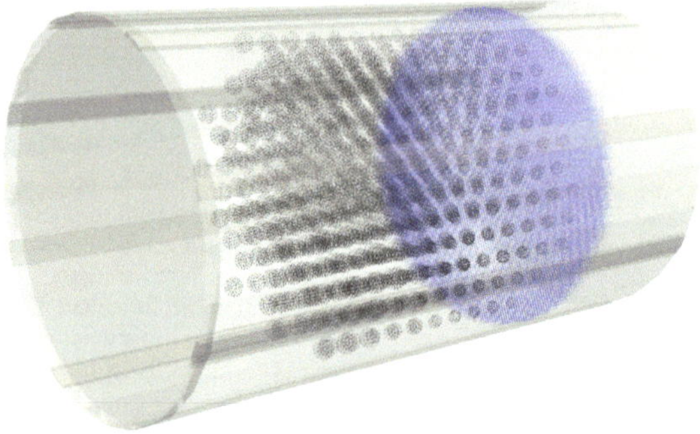

Figure 15. Geometry generation with balls and liquid generation.

The geometry is defined through a range of points creating a bounded area and specific shapes like cylinders and boxes to represent the physical environment and objects within the simulation. The point reference serves as the origin, while 'pointmin' and 'pointmax' define the extents of the simulation domain. Objects are created using commands like 'drawcylinder' and 'drawbox', with various attributes such as radius, size, and orientation (through rotation angles) to model the simulation environment accurately.

Motion is applied to objects using the '<motion>' section, specifying the type of movement (rotational and rectilinear), its duration, speed, and direction. For instance, rotational movement is defined with an angular velocity around a specified axis, while rectilinear motion specifies velocity in the XYZ coordinates. These motions can be applied to different objects, such as a rotating drum or a vertically moving piston, to simulate dynamic interactions within the fluid. Appropriate rotational speed was applied for each case to reflect the real rotational speed from experiments. An example of the movement of objects inside the mill is shown in Figure 16.

Figure 16. Movement of the mill wall like a piston.

3.8.2. Data Extraction

After each simulation, data about ball positions were extracted (Figure 17 (left)) for the sampled time steps. Due to the use of the same lifter heights for all diameters, for the smallest mill diameter, there was a single ball lifter higher than the others on dropped down; this produced the balls seen in the air. For that reason, the heatmap (Figure 17 (right)) showing the most frequent position of the ball charge was created to decrease the impact of those lifted balls on the overall measurement of the angle. This phenomenon occurred in the recordings and simulation, so the same further analysis flow was used for comparable results.

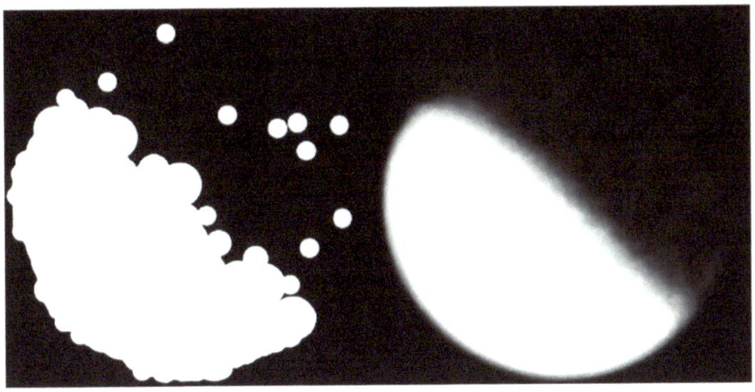

Figure 17. Spheres position in the simulation (example mill 300 mm) extracted and saved as binary image (on the left) heatmap generated from the images (on the right).

Additionally, for wet milling simulations, additional data about water elevation were extracted. The data were extracted from the plane in the middle of the mill parallel to the front and back walls. The data represented the highest coordinate vertical for each horizontal coordinate on the plane. Due to the discretization of the liquid and representing it with low resolution, there were a lot of outliers. To clean the data and make them more uniform, and reduce the recording of the splashed liquid into the air or the lack of liquid height, because of the low resolution and space occupied in the plane by the steel balls, for each horizontal coordinate (x), the medial position and Q3 were calculated, and only points between those values were taken into account in further curve fitting. Similar to the recording quadratic equation curve approach, a 95% confidence interval was calculated. Figure 18 shows the final load on the mill space after the data extraction process.

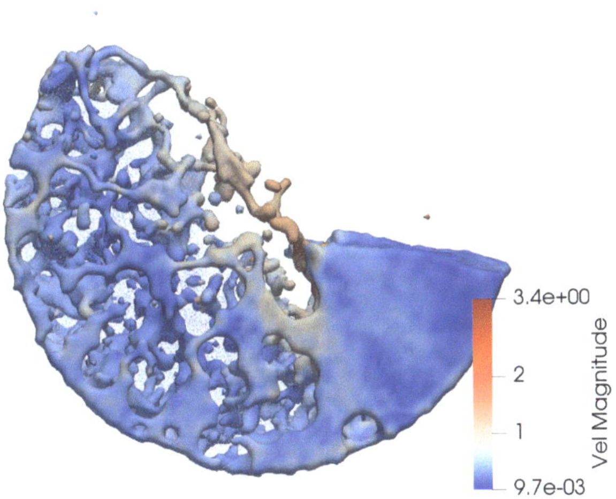

Figure 18. Load in the final space in the mill in simulation, vel (m/s).

3.9. Calibration

For both cases without liquid, only balls with a non-iterative approach were chosen. The behavior of DEM bodies (floats) was modified by changing two parameters: the coefficient of restitution and the coefficient of kinetic friction. Due to the instability of the simulation with higher values, the friction coefficient was explored in the range of 0.3–0.7 and the coefficient of restitution in the range of 0.3–0.55. For each iteration, three simulations with three different mill diameters were performed, and the obtained angles of repose were compared to the ones extracted from the image processing. The sum of square errors (SES) was recorded for each iteration. The ball calibration data from the first six iterations and coefficients and SSE combinations were used to interpolate values between the points. Then, the calibration algorithm was such that the following combination of coefficients tested would be a combination of the coefficients that gave the lowest SSE in the interpolated space. The algorithm was stopped when the three following combinations did not provide a better outcome than the previous combination.

The entire sets of data thus measured angles from simulations from ball calibration were still valuable for calibrating the balls with ore because the ore was not simulated separately; the contact parameters were planned to adjust for the different load behaviors represented by the steel balls. All obtained angles were compared to the ones measured from dry milling recordings to obtain the initial interpolated response surface. The algorithm was later the same. We explored combinations that lowered SSE until three subsequent tests did not give the best result.

As mentioned previously, wet milling recordings were taken into account only from the biggest diameter, so wet milling simulations were also performed only on the 500 mm diameter. Calibration of the liquid also required adjusting kernel parameters or viscosity. The behavior of the liquid is satisfying if both curves fit to the liquid surface for the recording and the simulation fits inside each other's 95% confidence intervals.

4. Results

The measured dynamic angles of repose for recordings of the ball inside the drum are 43.25, 42.15, and 40.35 degrees for 300, 400, and 500 mm, respectively. The dynamic angles recorded during dry milling processes were found to be 54.24 degrees for a 300 mm measurement, 41.97 degrees for 400 mm, and 47.67 degrees for 500 mm, respectively. Those values are less accurate due to the more challenging extraction of the area of the load in the recording and the reflections on the wall, which this method was not resistant to. The curve fitted to the data from the wet milling recording is presented below, in Figure 19.

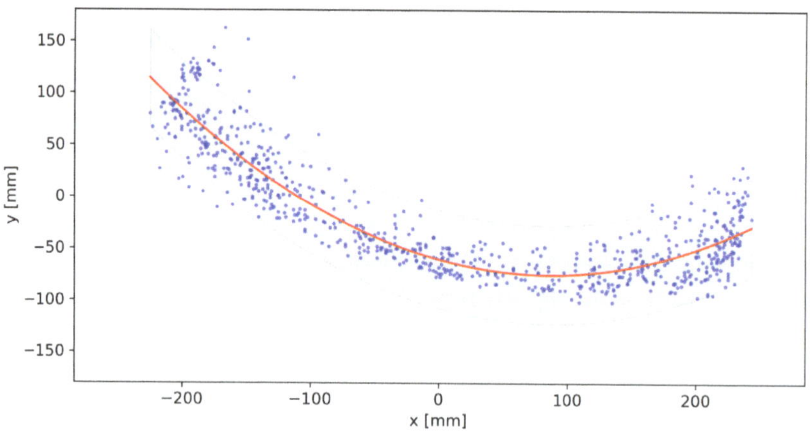

Figure 19. Points of the slurry surface detected with the GUI, fitted curve, and 95% confidence interval.

Calibration of the ball behavior started with six initial measurements of angles for each diameter in the simulation (3 × 6 = 18 initial simulations). New local minimums were discovered until the 10th iteration, after which no new better combination was found. The final parameters for simulation balls were coefficient of restitution = 0.390 and coefficient of kinetic friction = 0.496. In total, 13 iterations resulting in 39 simulations were performed to achieve those results. The results of the ball behavior calibration, as illustrated in Figure 20, are presented through heatmaps depicting the calibration process of ball behavior in a milling environment. These maps elucidate the relationship between two critical parameters: the coefficient of restitution and the coefficient of kinetic friction, across a color gradient that represents the sum of squared errors (SSE). Specifically, the x-axis details the coefficient of restitution, which quantifies the elasticity of collisions, and the y-axis corresponds to the coefficient of kinetic friction, which measures the resistance to motion. Contour lines represent constant SSE values, with the color gradient from green to red indicating ascending SSE values, thus reflecting the fidelity of the simulation parameters to the experimental observations. The optimal parameters are marked and exhibit the lowest SSE, pointing to the most accurate simulation values determined through the iterative calibration process.

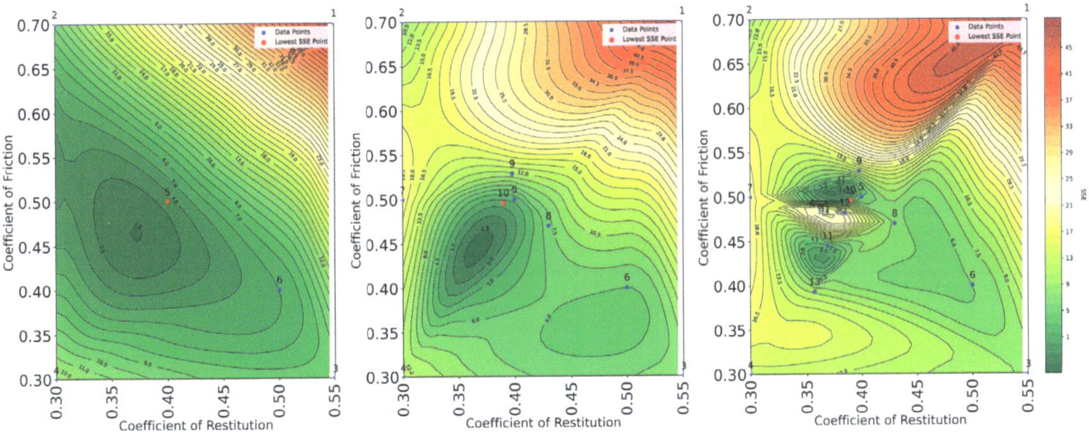

Figure 20. Calibration of balls behavior.

For the calibration of the dry milling, all resulting angles were considered to calculate the initial response surface with the new SSE value. In this case, the 12th iteration from the previous calibration resulted in the best result and lowest SSE. Three possible local minima were explored; none gave better results than the previous best combination from the 12th iteration. Thus, the selected parameters were as follows: coefficient of restitution 0.385 and coefficient of kinetic friction 0.482. Those parameters are less accurate because the recording's measurement of the load's angle of repose is less precise. The dry milling calibration results are shown in Figure 21, similar to the previous case, as a heatmap.

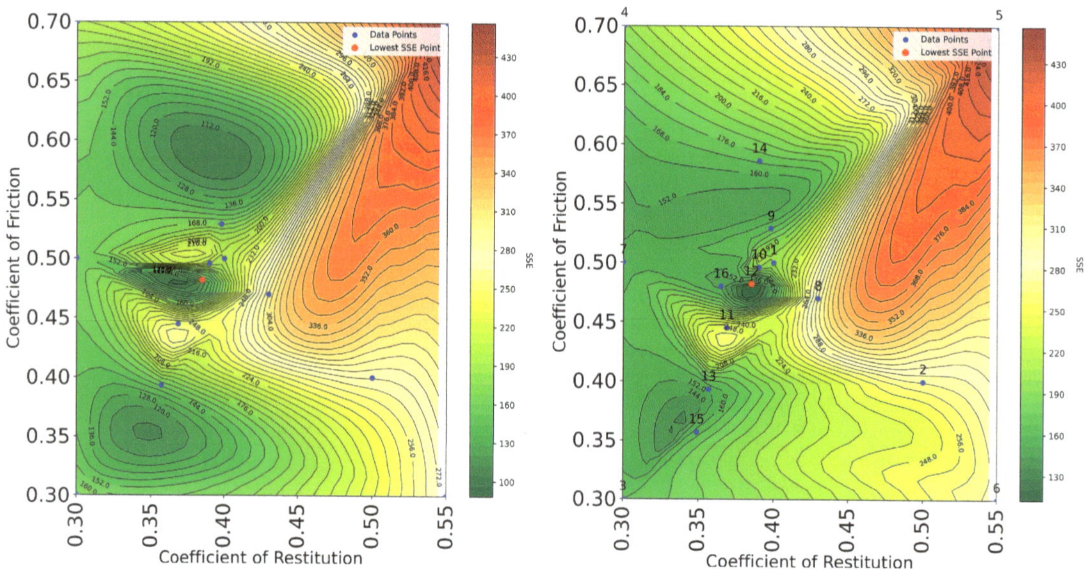

Figure 21. Calibration of dry milling.

The extracted water elevation points for all timesteps, cleaned as described before, are presented in Figure 22.

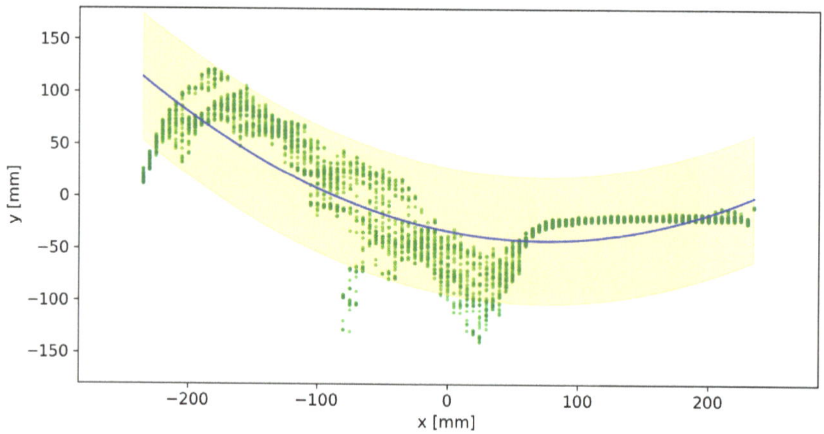

Figure 22. Points of the slurry surface extracted from the simulation, with fitted curve and 95% confidence interval.

When comparing these data to the data extracted from the recording (Figure 23), we can see that the fitted curves fit inside each other's confidence intervals, thus signifying that by just adjusting the fluid's density and viscosity to realistic values, we obtained a nice match in the behavior of the simulated slurry.

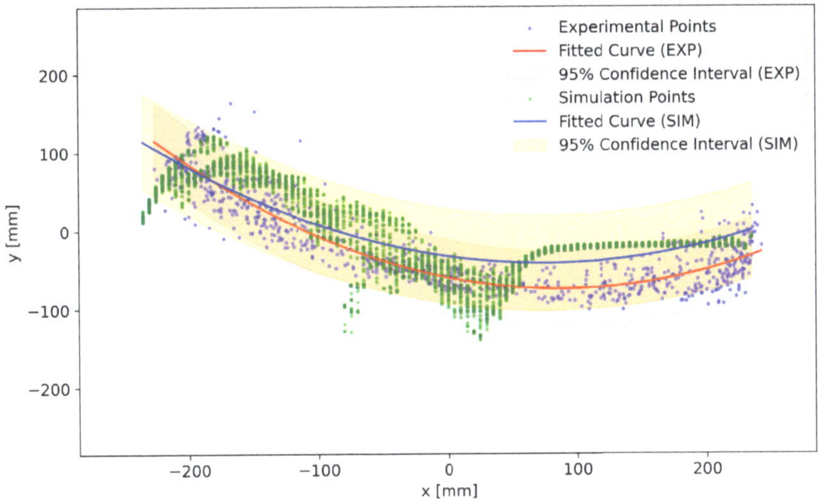

Figure 23. Comparison of the data from recording and simulation for wet milling calibration.

5. Discussion

This section delves into our study's sophisticated methods and significant insights, spotlighting the role of integrating cutting-edge computer vision with DEM-SPH techniques in enhancing the calibration of digital twins for lab ball mills. We aim to highlight the breakthroughs and future possibilities that this research opens up.

5.1. Embracing Advanced Computer Vision

Our work takes a giant leap forward by incorporating the latest computer vision technology, such as Meta's SAM model, to automate data gathering. This innovation allows us to precisely track and analyze how the ball mill operates under different conditions by looking at the dynamic angle of repose and monitoring the slurry surface. These detailed observations are crucial for fine-tuning the digital twin of the mill. However, we also encountered some hurdles, especially with the limited visibility in wet milling scenarios, pointing to a need for refining these techniques or finding new ways to gather data on such complex systems.

5.2. Advancing DEM-SPH Calibration

In refining the calibration process, we integrated DEM simulations for the grinding media with SPH models for the slurry flow, showcasing the challenge of capturing milling actions accurately. The careful adjustments of the restitution and friction coefficients in the calibration process underscore the importance of these parameters in mimicking the mill's real-world behavior. Our efforts in fine-tuning these aspects demonstrate the potential of this method in improving simulation precision, especially for copper ore milling, despite the limitations in measuring dry milling conditions and the promising yet challenging calibration of wet milling.

5.3. The Road Ahead for Digital Twin Calibration

Future work should enhance these calibration techniques to encapsulate the full complexity of industrial milling operations under various conditions. This might include new data collection methods and applying advanced technologies like AI to fine-tune model parameters. Achieving a robust calibration is crucial for harnessing the full predictive power of this approach.

5.4. Leveraging Digital Twins for Milling Efficiency

Our research underscores the transformative potential of calibrated digital twins in milling optimization, particularly for copper ore processing. By accurately simulating milling dynamics, we can explore how different operational settings affect efficiency and energy consumption without requiring extensive physical experiments. This opens up possibilities for designing more energy-efficient milling processes, marking a significant step forward for the mineral processing industry.

5.5. Exploring New Research Avenues

The findings also point to several promising areas for future research. Enhancing computer vision for better data accuracy in wet milling, understanding the long-term benefits of digital twin integration on process efficiency, equipment lifespan, and energy consumption, and conducting further experimental and simulation studies to explore the effects of milling parameters are all critical for realizing the full potential of this innovative approach. This work lays a solid foundation for the ongoing refinement of, and significant advancements in, process optimization.

6. Conclusions

This study presents a comprehensive methodology for calibrating the digital twin of a laboratory ball mill, emphasizing the milling of copper ore through the integration of advanced computer vision techniques and the coupling of DEM (Discrete Element Method) with SPH (Smoothed Particle Hydrodynamics) simulations. The calibration process involved meticulously examining the mill's behavior under various operational scenarios, including dry and wet milling, facilitated by state-of-the-art image processing and the robust DualSPHysics framework. The following conclusions can be drawn from the research conducted:

- **Enhanced Calibration Precision**: Integrating computer vision for data extraction and applying DEM-SPH simulations for various milling scenarios significantly improved the accuracy of the digital twin's behavior compared to actual mill operations. This enhanced precision is pivotal for optimizing milling strategies, leading to more efficient processing and energy utilization.
- **Optimization of Milling Processes**: The calibrated digital twin provides a powerful tool for understanding and optimizing the grinding process. By accurately simulating the dynamics of the grinding media and slurry, the study paves the way for developing milling processes that are energy efficient and capable of producing high-quality milled products. This optimization potential extends beyond copper ore milling to other materials and milling environments.
- **Future Research and Applications**: While the current study focuses on copper ore milling, the methodology developed has broader applications. Future research can extend this approach to other milling operations, materials, and industrial processes. Additionally, the integration of further advancements in computer vision and machine learning could automate and refine the calibration process, making digital twins even more accurate and valuable across various sectors.
- **Contribution to Sustainable Practices**: By enabling the optimization of milling operations for energy efficiency and material usage, the study contributes to more sustainable industrial practices. Simulating and predicting the outcomes of various operational parameters without the need for extensive physical testing conserves resources reduces the environmental footprint of mineral processing industries.

In conclusion, this study successfully demonstrates the viability and benefits of calibrating a digital twin for a laboratory ball mill using the novel integration of computer vision and DEM-SPH simulations. The insights gained and the methodology developed contribute significantly to mineral processing and digital twin technology, offering a promising avenue for future advancements in industrial milling operations and beyond. The calibrated digital twin is a testament to the potential of combining traditional engineering with cutting-edge technology to solve complex challenges in mineral processing and material science.

Author Contributions: B.D.: conceptualization, methodology, software, validation, formal analysis, investigation, data curation, writing—original draft preparation, visualization; project administration, funding acquisition; P.B.: conceptualization, methodology, validation, formal analysis, data curation, investigation, original draft preparation; M.O.: conceptualization, validation, formal analysis, investigation, data curation, writing—original draft preparation; R.K.: resources, supervision, funding acquisition, writing—review and editing. All authors have read and agreed to the published version of the manuscripts.

Funding: Scientific work funded from the state budget for science in the years 2019–2024, as a research project under the "Diamond Grant" (pl. "Diamentowy Grant") program and from the subsidy of the Minister of Education for 2024.

Institutional Review Board Statement: Not applicable.

Informed Consent Statement: Not applicable.

Data Availability Statement: The data are available upon email contact with the corresponding author.

Conflicts of Interest: The authors declare no conflicts of interest.

References

1. Farjana, S.H.; Huda, N.; Parvez Mahmud, M.; Saidur, R. A review on the impact of mining and mineral processing industries through life cycle assessment. *J. Clean. Prod.* **2019**, *231*, 1200–1217. [CrossRef]
2. Cleary, P.W.; Morrison, R.D.; Sinnott, M.D. Prediction of slurry grinding due to media and coarse rock interactions in a 3D pilot SAG mill using a coupled DEM + SPH model. *Miner. Eng.* **2020**, *159*, 106614. [CrossRef]
3. Al-Thyabat, S.; Nakamura, T.; Shibata, E.; Iizuka, A. Adaptation of minerals processing operations for lithium-ion (LiBs) and nickel metal hydride (NiMH) batteries recycling: Critical review. *Miner. Eng.* **2013**, *45*, 4–17. [CrossRef]
4. Tohry, A.; Chehreh Chelgani, S.; Matin, S.; Noormohammadi, M. Power-draw prediction by random forest based on operating parameters for an industrial ball mill. *Adv. Powder Technol.* **2020**, *31*, 967–972. [CrossRef]
5. Fuerstenau, D.; Abouzeid, A.Z. The energy efficiency of ball milling in comminution. *Int. J. Miner. Process.* **2002**, *67*, 161–185. [CrossRef]
6. Sverak, T.; Baker, C.; Kozdas, O. Efficiency of grinding stabilizers in cement clinker processing. *Miner. Eng.* **2013**, *43-44*, 52–57. [CrossRef]
7. Bu, X.; Chen, Y.; Ma, G.; Sun, Y.; Ni, C.; Xie, G. Wet and dry grinding of coal in a laboratory-scale ball mill: Particle-size distributions. *Powder Technol.* **2020**, *359*, 305–313. [CrossRef]
8. Shi, F.; Xie, W. A specific energy-based size reduction model for batch grinding ball mill. *Miner. Eng.* **2015**, *70*, 130–140. [CrossRef]
9. Ogonowski, S.; Wołosiewicz-Głąb, M.; Ogonowski, Z.; Foszcz, D.; Pawełczyk, M. Comparison of Wet and Dry Grinding in Electromagnetic Mill. *Minerals* **2018**, *8*, 138. [CrossRef]
10. He, M.; Wang, Y.; Forssberg, E. Slurry rheology in wet ultrafine grinding of industrial minerals: A review. *Powder Technol.* **2004**, *147*, 94–112. [CrossRef]
11. Breitung-Faes, S.; Kwade, A. Mill, material, and process parameters—A mechanistic model for the set-up of wet-stirred media milling processes. *Adv. Powder Technol.* **2019**, *30*, 1425–1433. [CrossRef]
12. Chelgani, S.C.; Parian, M.; Parapari, P.S.; Ghorbani, Y.; Rosenkranz, J. A comparative study on the effects of dry and wet grinding on mineral flotation separation–a review. *J. Mater. Res. Technol.* **2019**, *8*, 5004–5011. [CrossRef]
13. Kotake, N.; Kuboki, M.; Kiya, S.; Kanda, Y. Influence of dry and wet grinding conditions on fineness and shape of particle size distribution of product in a ball mill. *Adv. Powder Technol.* **2011**, *22*, 86–92. [CrossRef]
14. Cleary, P.W.; Morrison, R.D. Prediction of 3D slurry flow within the grinding chamber and discharge from a pilot scale SAG mill. *Miner. Eng.* **2012**, *39*, 184–195. [CrossRef]
15. Morrell, S. Modelling the influence on power draw of the slurry phase in Autogenous (AG), Semi-autogenous (SAG) and ball mills. *Miner. Eng.* **2016**, *89*, 148–156. [CrossRef]
16. Sinnott, M.; Cleary, P.; Morrison, R. Combined DEM and SPH simulation of overflow ball mill discharge and trommel flow. *Miner. Eng.* **2017**, *108*, 93–108. [CrossRef]
17. Mayank, K.; Malahe, M.; Govender, I.; Mangadoddy, N. Coupled DEM-CFD Model to Predict the Tumbling Mill Dynamics. *Procedia IUTAM* **2015**, *15*, 139–149. [CrossRef]
18. Soleymani, M.M.; Fooladi, M.; Rezaeizadeh, M. Effect of slurry pool formation on the load orientation, power draw, and impact force in tumbling mills. *Powder Technol.* **2016**, *287*, 160–168. [CrossRef]
19. Yin, Z.; Peng, Y.; Zhu, Z.; Ma, C.; Yu, Z.; Wu, G. Effect of mill speed and slurry filling on the charge dynamics by an instrumented ball. *Adv. Powder Technol.* **2019**, *30*, 1611–1616. [CrossRef]

20. Yuan, C.; Wu, C.; Fang, X.; Liao, N.; Tong, J.; Yu, C. Effect of Slurry Concentration on the Ceramic Ball Grinding Characteristics of Magnetite. *Minerals* **2022**, *12*, 1569. [CrossRef]
21. Xie, C.; Ma, H.; Song, T.; Zhao, Y. DEM investigation of SAG mill with spherical grinding media and non-spherical ore based on polyhedron-sphere contact model. *Powder Technol.* **2021**, *386*, 154–165. [CrossRef]
22. Dubé, O.; Alizadeh, E.; Chaouki, J.; Bertrand, F. Dynamics of non-spherical particles in a rotating drum. *Chem. Eng. Sci.* **2013**, *101*, 486–502. [CrossRef]
23. Tang, J.; Yu, W.; Chai, T.; Liu, Z.; Zhou, X. Selective ensemble modeling load parameters of ball mill based on multi-scale frequency spectral features and sphere criterion. *Mech. Syst. Signal Process.* **2016**, *66–67*, 485–504. [CrossRef]
24. Senapati, P.; Mishra, B.; Parida, A. Modeling of viscosity for power plant ash slurry at higher concentrations: Effect of solids volume fraction, particle size and hydrodynamic interactions. *Powder Technol.* **2010**, *197*, 1–8. [CrossRef]
25. Roco, M.; Shook, C. Modeling of slurry flow: The effect of particle size. *Can. J. Chem. Eng.* **1983**, *61*, 494–503. [CrossRef]
26. Lakhdissi, E.M.; Fallahi, A.; Guy, C.; Chaouki, J. Effect of solid particles on the volumetric gas liquid mass transfer coefficient in slurry bubble column reactors. *Chem. Eng. Sci.* **2020**, *227*, 115912. [CrossRef]
27. Zhang, X.; Ahmadi, G. Eulerian–Lagrangian simulations of liquid-gas-solid flows in three-phase slurry reactors. *Chem. Eng. Sci.* **2005**, *60*, 5089–5104. [CrossRef]
28. Buszko, M.H.; Krella, A.K. An Influence of Factors of Flow Condition, Particle and Material Properties on Slurry Erosion Resistance. *Adv. Mater. Sci.* **2019**, *19*, 28–53. [CrossRef]
29. Olhero, S.; Ferreira, J. Influence of particle size distribution on rheology and particle packing of silica-based suspensions. *Powder Technol.* **2004**, *139*, 69–75. [CrossRef]
30. Zhang, J.; Zhao, H.; Wang, C.; Li, W.; Xu, J.; Liu, H. The influence of pre-absorbing water in coal on the viscosity of coal water slurry. *Fuel* **2016**, *177*, 19–27. [CrossRef]
31. Senapati, P.K.; Panda, D.; Parida, A. Predicting viscosity of limestone–water slurry. *J. Miner. Mater. Charact. Eng.* **2009**, *8*, 203. [CrossRef]
32. Gillies, R.G.; Shook, C.A. Modelling high concentration settling slurry flows. *Can. J. Chem. Eng.* **2000**, *78*, 709–716. [CrossRef]
33. Andersson, V.; Gudmundsson, J.S. Flow properties of hydrate-in-water slurries. *Ann. N. Y. Acad. Sci.* **2000**, *912*, 322–329. [CrossRef]
34. Domínguez, J.M.; Fourtakas, G.; Altomare, C.; Canelas, R.B.; Tafuni, A.; García-Feal, O.; Martínez-Estévez, I.; Mokos, A.; Vacondio, R.; Crespo, A.J.C.; et al. DualSPHysics: From fluid dynamics to multiphysics problems. *Comput. Part. Mech.* **2021**, *9*, 867–895. [CrossRef]
35. Shadloo, M.; Oger, G.; Le Touzé, D. Smoothed particle hydrodynamics method for fluid flows, towards industrial applications: Motivations, current state, and challenges. *Comput. Fluids* **2016**, *136*, 11–34. [CrossRef]
36. Gotoh, H.; Khayyer, A. On the state-of-the-art of particle methods for coastal and ocean engineering. *Coast. Eng. J.* **2018**, *60*, 79–103. [CrossRef]
37. Mogan, S.C.; Chen, D.; Hartwig, J.; Sahin, I.; Tafuni, A. Hydrodynamic analysis and optimization of the Titan submarine via the SPH and Finite—Volume methods. *Comput. Fluids* **2018**, *174*, 271–282. [CrossRef]
38. Manenti, S.; Wang, D.; Domínguez, J.M.; Li, S.; Amicarelli, A.; Albano, R. SPH Modeling of Water-Related Natural Hazards. *Water* **2019**, *11*, 1875. [CrossRef]
39. Domínguez, J.M.; Crespo, A.J.C.; Gómez-Gesteira, M.; Marongiu, J.C. Neighbour lists in smoothed particle hydrodynamics. *Int. J. Numer. Methods Fluids* **2011**, *67*, 2026–2042. [CrossRef]
40. Domínguez, J.M.; Crespo, A.J.; Gómez-Gesteira, M. Optimization strategies for CPU and GPU implementations of a smoothed particle hydrodynamics method. *Comput. Phys. Commun.* **2013**, *184*, 617–627. [CrossRef]
41. Dehnen, W.; Aly, H. Improving convergence in smoothed particle hydrodynamics simulations without pairing instability. *Mon. Not. R. Astron. Soc.* **2012**, *425*, 1068–1082. [CrossRef]
42. Liu, M.; Liu, G.; Lam, K. Constructing smoothing functions in smoothed particle hydrodynamics with applications. *J. Comput. Appl. Math.* **2003**, *155*, 263–284. [CrossRef]
43. Monaghan, J.J. Smoothed Particle Hydrodynamics. *Annu. Rev. Astron. Astrophys.* **1992**, *30*, 543–574. [CrossRef]
44. Wendland, H. Piecewise polynomial, positive definite and compactly supported radial functions of minimal degree. *Adv. Comput. Math.* **1995**, *4*, 389–396. [CrossRef]
45. Robinson, M. Turbulence and Viscous Mixing Using Smoothed Particle Hydrodynamics. Ph.D. Thesis, Monash University, Clayton, Australia, 2010.
46. Monaghan, J. Smoothed particle hydrodynamics. *Rep. Prog. Phys.* **2005**, *68*, 1703–1759. [CrossRef]
47. Violeau, D. *Fluid Mechanics and the SPH Method*; Oxford University Press: Oxford, UK, 2012.
48. Violeau, D.; Rogers, B. Smoothed particle hydrodynamics (SPH) for free-surface flows: Past, present and future. *J. Hydraul. Res.* **2016**, *54*, 1–26. [CrossRef]
49. Antuono, M.; Colagrossi, A.; Marrone, S. Numerical diffusive terms in weakly-compressible SPH schemes. *Comput. Phys. Commun.* **2012**, *183*, 2570–2580. [CrossRef]
50. Bonet, J.; Lok, T.S. Variational and momentum preservation aspects of smooth particle hydrodynamic formulations. *Comput. Methods Appl. Mech. Eng.* **1999**, *180*, 97–115. [CrossRef]

51. Molteni, D.; Colagrossi, A. A simple procedure to improve the pressure evaluation in hydrodynamic context using the SPH. *Comput. Phys. Commun.* **2009**, *180*, 861–872. [CrossRef]
52. Fourtakas, G.; Dominguez, J.; Vacondio, R.; Rogers, B. Local uniform stencil (LUST) boundary condition for arbitrary 3-D boundaries in parallel smoothed particle hydrodynamics (SPH) models. *Comput. Fluids* **2019**, *190*, 346–361. [CrossRef]
53. Monaghan, J.; Gingold, R. Shock simulation by the particle method SPH. *J. Comput. Phys.* **1983**, *52*, 374–389. [CrossRef]
54. Español, P.; Revenga, M. Smoothed dissipative particle dynamics. *Phys. Rev. E* **2003**, *67*, 026705. [CrossRef] [PubMed]
55. Lo, Y.; Shao, S. Simulation of near-shore solitary wave mechanics by an incompressible SPH method. *Appl. Ocean. Res.* **2002**, *24*, 275–286. [CrossRef]
56. Colagrossi, A.; Antuono, M.; Souto-Iglesias, A.; Le Touzé, D. Theoretical analysis and numerical verification of the consistency of viscous smoothed-particle-hydrodynamics formulations in simulating free-surface flows. *Phys. Rev. E Stat. Nonlinear Soft Matter Phys.* **2011**, *84*, 026705. [CrossRef] [PubMed]
57. Gotoh, H.; Shibahara, T.; Sakai, T. Sub-particle-scale Turbulence Model for the MPS Method: Lagrangian flow model for hydraulic engineering. In *Advanced Methods for Computational Fluid Dynamics*; TIB: Hannover, Germany, 2001.
58. Dalrymple, R.; Rogers, B. Numerical modeling of water waves with the SPH method. *Coast. Eng.* **2006**, *53*, 141–147. [CrossRef]
59. Monaghan, J. Simulating free surface flows with SPH. *J. Comput. Phys.* **1994**, *110*, 399–406. [CrossRef]
60. Batchelor, G. *An Introduction to Fluid Dynamics*; Cambridge University Press: Cambridge, UK, 2000.
61. Verlet, L. Computer 'experiments' on classical fluids. I. Thermodynamical properties of Lennard-Jones molecules. *Phys. Rev.* **1967**, *159*, 98–103. [CrossRef]
62. Leimkuhler, B.; Matthews, C. Introduction In *Molecular Dynamics: With Deterministic and Stochastic Numerical Methods*; Springer: Berlin/Heidelberg, Germany, 2015; pp. 1–51.
63. Parshikov, A.; Medin, S.; Loukashenko, I.; Milekhin, V. Improvements in SPH method by means of interparticle contact algorithm and analysis of perforation tests at moderate projectile velocities. *Int. J. Impact Eng.* **2000**, *24*, 779–796. [CrossRef]
64. Monaghan, J.; Kos, A. Solitary waves on a Cretan beach. *J. Waterw. Port Coast. Ocean. Eng.* **1999**, *125*, 145–155. [CrossRef]
65. Altomare, C.; Crespo, A.; Rogers, B. Numerical modelling of armour block sea breakwater with smoothed particle hydrodynamics. *Comput. Struct.* **2014**, *130*, 34–45. [CrossRef]
66. Zhang, F.; Crespo, A.; Altomare, C. DualSPHysics: A numerical tool to simulate real breakwaters. *J. Hydrodyn.* **2018**, *30*, 95–105. [CrossRef]
67. English, A.; Domínguez, J.; Vacondio, R. Modified dynamic boundary conditions (mDBC) for general purpose smoothed particle hydrodynamics (SPH): Application to tank sloshing, dam break and fish pass problems. *Comput. Part. Mech.* **2021**, *9*, 1–5. [CrossRef]
68. Crespo, A.; Gómez-Gesteira, M.; Dalrymple, R. Boundary conditions generated by dynamic particles in SPH methods. *Comput. Mater. Contin.* **2007**, *5*, 173–184. [CrossRef]
69. Lind, S.; Xu, R.; Stansby, P.; Rogers, B. Incompressible smoothed particle hydrodynamics for free-surface flows: A generalised diffusion-based algorithm for stability and validations for impulsive flows and propagating waves. *J. Comput. Phys.* **2012**, *231*, 1499–1523. [CrossRef]
70. Skillen, A.; Lind, S.; Stansby, P.; Rogers, B. Incompressible smoothed particle hydrodynamics (SPH) with reduced temporal noise and generalised Fickian smoothing applied to body–water slam and efficient wave-body interaction. *Comput. Methods Appl. Mech. Eng.* **2013**, *265*, 163–173. [CrossRef]
71. Canelas, R.B.; Crespo, A.J.; Domínguez, J.M.; Ferreira, R.M.; Gómez-Gesteira, M. SPH-DCDEM model for arbitrary geometries in free surface solid-fluid flows. *Comput. Phys. Commun.* **2016**, *202*, 131–140. [CrossRef]
72. Brilliantov, N.; Pöschel, T. Granular gases with impact-velocity-dependent restitution coefficient. In *Granular Gases*; Springer: Berlin/Heidelberg, Germany, 2001; pp. 100–124. [CrossRef]
73. Cummins, S.; Cleary, P. Using distributed contacts in DEM. *Appl. Math. Model.* **2011**, *35*, 1904–1914. [CrossRef]
74. Colagrossi, A.; Landrini, M. Numerical simulation of interfacial flows by smoothed particle hydrodynamics. *J. Comput. Phys.* **2003**, *191*, 448–475. [CrossRef]
75. Koshizuka, S.; Nobe, A.; Oka, Y. Numerical analysis of breaking waves using the moving particle semi-implicit method. *Int. J. Numer. Methods Fluids* **1998**, *26*, 751–769. [CrossRef]
76. Canelas, R.B.; Domínguez, J.M.; Crespo, A.J.; Gómez-Gesteira, M.; Ferreira, R.M. A Smooth Particle Hydrodynamics discretization for the modelling of free surface flows and rigid body dynamics. *Int. J. Numer. Methods Fluids* **2015**, *78*, 581–593. [CrossRef]
77. Price, D.J. Modelling discontinuities and Kelvin–Helmholtz instabilities in SPH. *J. Comput. Phys.* **2008**, *227*, 10040–10057. [CrossRef]
78. Saitoh, T.R.; Makino, J. A density-independent formulation of smoothed particle hydrodynamics. *Astrophys. J.* **2013**, *768*, 44. [CrossRef]
79. Kuwabara, G.; Kono, K. Restitution coefficient in a collision between two spheres. *Jpn. J. Appl. Phys.* **1987**, *26*, 1230. [CrossRef]
80. Kruggel-Emden, H.; Simsek, E.; Rickelt, S.; Wirtz, S.; Scherer, V. Review and extension of normal force models for the discrete element method. *Powder Technol.* **2007**, *171*, 157–173. [CrossRef]
81. Lemieux, M.; Léonard, G.; Doucet, J.; Leclaire, L.A.; Viens, F.; Chaouki, J.; Bertrand, F. Large-scale numerical investigation of solids mixing in a V-blender using the discrete element method. *Powder Technol.* **2008**, *181*, 205–216. [CrossRef]
82. Vetsch, D. Numerical Simulation of Sediment Transport with Meshfree Methods. Ph.D. Thesis, ETH, Zurich, Switzerland, 2011.

83. Hoomans, B. Granular Dynamics of Gas-Solid Two-Phase Flows. Ph.D. Thesis, University of Twente, Enschede, The Netherlands, 2000.
84. Francis, W.; Peters, M.C. Data Sheet No.101-Classification, Properties and Units. In *Fuels and Fuel Technology*, 2nd ed.; Springer: Berlin/Heidelberg, Germany, 1980; pp. 313–319. [CrossRef]
85. Dunne, R.M.E. *SME Mineral Processing & Extractive Metallurgy Handbook*; Society for Mining, Metallurgy & Exploration (SME): Englewood, CO, USA, 2019.
86. Bond, F. *Crushing and Grinding Calculations*; Allis-Chalmers Manufacturing: Milwaukee, WI, USA, 1961.
87. Gupta, A.; Yan, D.S. *Mineral Processing Design and Operations: An Introduction*, 2nd ed.; Elsevier: Amsterdam, The Netherlands, 2016.
88. Kirillov, A.; Mintun, E.; Ravi, N.; Mao, H.; Rolland, C.; Gustafson, L.; Xiao, T.; Whitehead, S.; Berg, A.C.; Lo, W.Y.; et al. Segment Anything. *arXiv* **2023**, arXiv:2304.02643.

minerals

MDPI

Article

Time-Dependent Tap Density Modeling of Graphite Milled by Vibrating Disc Mill

Gülşah Güven [1], Ugur Ulusoy [2,*], Fırat Burat [1], Behrad Mojtahedi [3] and Guler Bayar [4]

1 Mineral Processing Engineering Department, Istanbul Technical University, TR34467 Istanbul, Turkey; guveng15@itu.edu.tr (G.G.); buratf@itu.edu.tr (F.B.)
2 Chemical Engineering Department, Sivas Cumhuriyet University, TR58140 Sivas, Turkey
3 Materials Science and Engineering, Sharif University of Technology, Tehran 1458889694, Iran; behrad.mojtahedi25@sharif.edu
4 Nanotechnology Engineering Department, Sivas Cumhuriyet University, TR58140 Sivas, Turkey; gbayar@cumhuriyet.edu.tr
* Correspondence: uulusoy@cumhuriyet.edu.tr

Abstract: Graphite, which is a key anode material for LIB, needs to have a high tap density (d_t) to reach a high volumetric energy density. Since d_t is directly correlated with particle size, particle size distribution, and particle shape, it can usually be improved by optimized grinding. So, determining the ideal grinding time by modeling the change in d_t over grinding time can yield substantial benefits like time, energy, and economy. However, the grinding time-dependent d_t modeling of graphite has never been reported before. Therefore, in this study, the relationship between the measured d_t values and grinding times of graphite particles by a vibrating disc mill (*VDM*) was investigated. Then, the empirical time-dependent d_t models were established with high R^2 values. The experimental and predicted d_t values were found to be close to each other. Among all tested fitting models, the exponential model ($d_t = ae^{-bt}$) was found to be the best-fitting model, having the highest R^2 and lowest error values. This approach provides guidance in the powder flow and processing of ground mineral materials, in the preparation processes of high-density graphite LIB anode material, as well as in graphite grinding in other mills in the industry, as well as in different electrode materials.

Keywords: tap density; graphite; modeling; degree of sphericity; vibratory disc mill; LIB anode

1. Introduction

Since the development of energy storage technologies for the transportation industry is of great importance, considering fluctuating oil costs, concerns about the negative impact of hydrocarbon emissions on global climate change, and energy security, modern batteries, which power everything from electric cars to mobile phones, have completely transformed our lives. In other words, lithium-ion batteries (*LIBs*), the most widely used battery chemistry for electric vehicles, personal electronics, and large-scale (grid) storage, are leading the current global shift towards renewable energy [1]. Therefore, research has focused on increasing the capacity and energy density of *LIBs* for use in hybrid electric vehicles (*HEVs*), plug-in HEVs, and fully electric vehicles.

However, characterizing and improving batteries for increased longevity and performance continues to present challenges. This calls for a more thorough comprehension of the characteristics of battery materials under progressively harsher operating circumstances [2,3]. Run time, safety, cycle life, power, energy density, and cost are the main performance characteristics of a battery that are determined by the materials that make up the cathode, anode, separator, and electrolyte. Particularly, a battery's volumetric energy density is an important factor to consider. Every day, researchers are doing extensive work to increase the energy density and power in *LIBs* since energy storage capacity is crucial for the *LIB* market. Because *LIB* is thought to be the best option as it offers the highest energy density of practical rechargeable batteries [4].

In this context, although there are many alternatives to *LIB* cathode material, graphite is the most widely used anode active material today and is, therefore, indispensable [3,5]. For example, a Tesla Model S, the most common *EV*, is known to use around 56 kg of graphite [1]. In addition, graphite is one of the critical raw materials that developed countries such as China, the EU, and the USA care about [6]. The growing emphasis on sustainability and environmental impact is expected to lead to a rise in the use of natural graphite because of its long cycle life, high energy density, and power density. Particularly, it has a long lifespan because of its layered crystal structure, which is perfect for the reversible intercalation of lithium ions and excellent electrical conductivity [1].

High specific capacity and high tap density (d_t) electrode materials are required for *LIBs* with high power densities [7]. In this context, d_t, which is a significant physical property of *LIB* anode material and influences the energy density of a *LIB* [4], is crucial insofar as it increases the electrodes' packing density to store more energy with the same volume [8]. Therefore, increasing the sphericity and d_t of the graphite particles is important for improving their volumetric energy density [9]. Thus, it can be used to provide a deeper comprehension of the physical characteristics of powder samples in practical applications to assess powder quality and product performance. When evaluating a powder's flow characteristics, the d_t assessment is essential since it provides important information about how the powder will be handled and processed [10]. In other words, the importance of high-d_t construction for *LIB* electrode materials has grown in recent years. For instance, during *LIB* manufacturing, increasing the d_t can maximize the high energy density. Whether the powder material consists of granules or flakes changes the way the particles are packed together and influences their values of d_t. While tapping, the particles are increasingly packed more efficiently, reducing the powder volume and ultimately increasing the d_t [11]. In fact, it can be increased by precisely optimizing several parameters, such as precursor preparation, calcination, and grinding [7,12]. The grinding performance of a mill depends on some parameters: energy consumption, efficiency, grinding size, and particle size distribution are all significantly impacted by the selection of additional grinding parameters. As a result, the fracture modes used by various mill types will influence the morphology of the particles. Thus, grinding with a suitable mill for sufficient time becomes very important as it accounts for 4% of global energy use [13].

In general, the d_t parameter is considered to be largely related to particle size, particle size distribution, and particle shape [7,8,12,14]. Since it has been reported that improving the graphite particles' volumetric energy density requires increasing d_t [15], determination of the optimal and achievable d_t of graphite material to produce a high volumetric energy density of *LIB* anode material by modeling and investigating the effect of grinding time on d_t values can address this issue [7], i.e., if time-dependent d_t models are used to estimate the optimal grinding time, this can help produce graphite particles with a greater d_t for anode usage. Thus, the rate at which the d_t value of the particles produced by grinding changes is useful knowledge for forecasting the grinding time needed, selecting an appropriate mill, and forecasting how ground graphite would behave in a subsequent operation. In other words, this approach can be used to calculate the necessary grinding time for the desired d_t value of graphite particles using a more suitable mill for the intended usage, considering energy and cost savings during milling.

Although it is expected that d_t can be improved by optimization of the production process, such as grinding [7], no attempts were made to investigate the rate of change in the d_t of ground graphite particles by a mill with various grinding times. Previous research [16] showed that although the vibrating disc mill (*VDM*) has limited capacity, it has higher cost and energy efficiency advantages than the ball mill (*BM*). Since the energy consumption in ultra-fine micronized grinding is very high, the VDM has proven to be an economical device for efficient ultra-fine micronized grinding [17]. Moreover, the *VDM* grinds faster than BM, and it was suggested that the *VDM* can produce good results by time-based shape models [16]. Therefore, *VDM* was selected as the grinding process of this study, which investigates the d_t changes in graphite particles as a function of grinding time.

With this study, it will be possible to intentionally adjust the attributes to match the needs of certain applications by providing fine control over the interaction between the d_t of graphite particles and grinding times. Thus, the best-desired results can be obtained by optimizing the grinding to select a suitable grinding system and grinding time. In other words, the aim of this study is to establish empirical time-dependent models between the average d_t of ground graphite particles and grinding time for the prediction of d_t values.

2. Materials and Methods

2.1. Sample Preparation

Medium-purity (72% C) graphite ore prepared by crushing and screening to a size fraction of 0.600–0.850 mm from the Oysu region of Kütahya, Turkey, was used for this research. When interpreted according to the X-ray diffraction (*XRD)* (Figure 1a) pattern and the X-ray spectroscopy (*EDS)* spectrum in scanning electron microscopy (*SEM)* given in Figure 1b, it is known that graphite is the major mineral and calcite, quartz, hematite, and clay group minerals are minor minerals in the ore. In this study, a laboratory-type vibratory disc mill (*VDM*) (Ünal Mühendislik, Turkey) with a diameter of 18.8 cm, a height of 6 cm, a volume of 500 mL, and rings with diameters of 7.4, 9.4, and 14 cm was used (Figure 1c). Sample preparation and grinding tests of graphite ore with a *VDM* have already been described in our previous study [16]. After the controlled crushing of the lump ore with a jaw crusher, the feedstock material of one size fraction was prepared for the grinding test with 600- and 850-micron sieves. The products obtained at the end of grinding at 10, 20, 30, 40, 50, and 60 s were separated into two narrow size fractions (0.150–0.250 mm and 0.053–0.150 mm) with 250-, 150-, and 53-micron sieves to minimize the size effect and were used in d_t measurements after sizing.

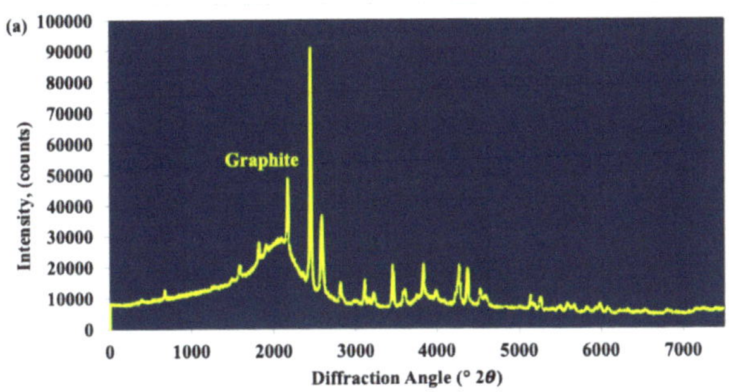

Figure 1. *Cont.*

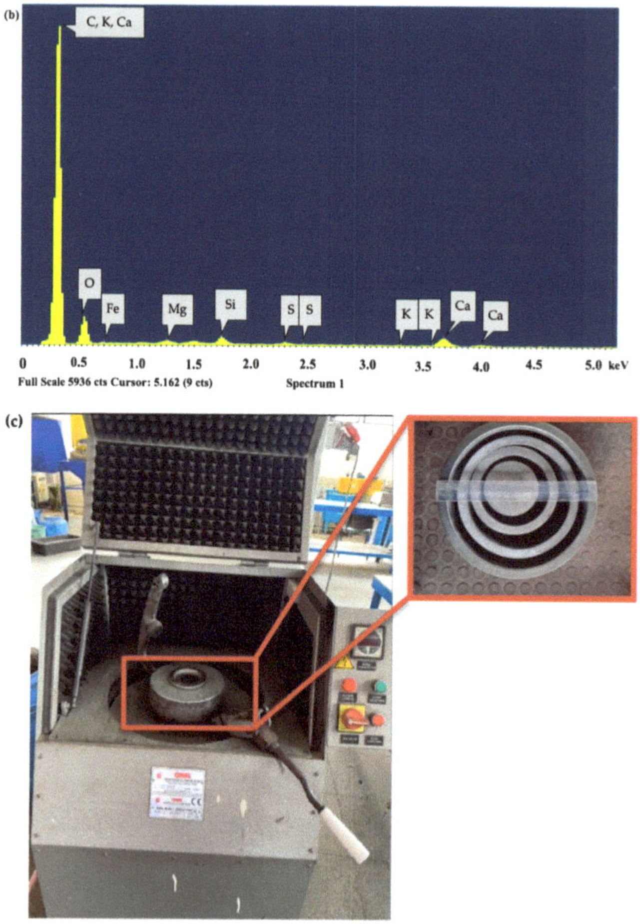

Figure 1. Graphite sample used in this study: (**a**) *XRD* pattern, (**b**) *SEM* spectrum, (**c**) VDM and rings used in the study.

2.2. d_t Measurements

Determination of d_t of graphite powder samples ground in a *VDM* for various grinding times and prepared in both size fractions was carried out using a d_t apparatus (BeDensi T2, Bettersize Instruments Ltd., Liaoning, China) as shown in Figure 2, which has a 1.0% repeatability, complying with ISO 3953:2011 standard [18]. Approximately 10 g of graphite sample was poured into the 25 cm^3 graduated glass cylinder for each test. The BeDensi T2 device was operated at a frequency of 300 strokes/min and a total of 2000 strokes. After the tapping process was completed, the final volume was recorded so that the surface of the sample was flat. Finally, the powder's d_t is calculated by dividing its mass by its volume as given by Equation (1):

$$d_t = m/V \tag{1}$$

where d_t, m, and V are defined as tap density (g/cm^3), mass of the powder (g), and volume (cm^3) of the tap powder, respectively. Each experiment was repeated three times, and the d_t values were calculated using the arithmetic mean of the values obtained.

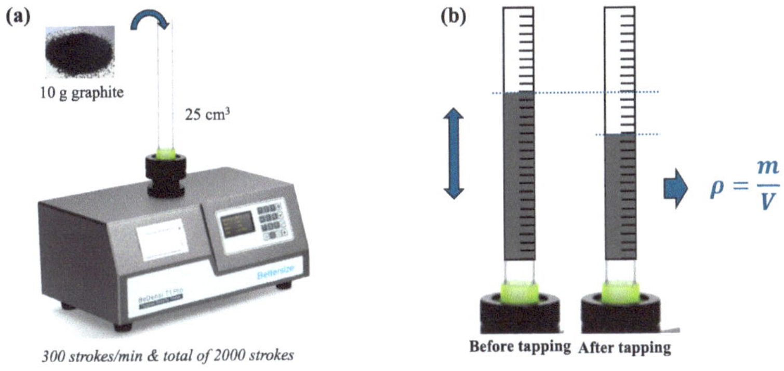

Figure 2. (**a**) d_t apparatus used in the study, (**b**) measurement of d_t.

2.3. Concept of Particle Degree of Sphericity

In this study, Dynamic Image Analysis (*DIA*) [19–22], which has been proven to be a technique that provides the fastest and most reliable results on the shape characterization of ground mineral particles according to previously published studies, was used. Shape characterization of the same graphite particles ground by a *VDM* with various grinding times was previously described [16] using *DIA* 1.0 (Micromeritics® Instrument Corp., Norcross, GA, USA).

According to the bubble–particle attachment test results in our last [23], which investigated the attachment efficiency of spherical, cylindrical, triangular prismatic, and cubic particles to air bubbles, according to attachment efficiency, settling speed, collision efficiency, and induction time, and investigated the role of particle geometry (surface, edges, and apex) on flotation behavior; since the attachment efficiency was shown to be in the order of cubic > triangular > cylindrical > spherical, in this study, we found it appropriate to define the tap density–shape relationship as a shape parameter (degree of sphericity) that looks more like a lumped cube instead of a rectangular or elongated particle.

As seen in Figure 3, various related particle shape metrics, such as aspect ratio (AR), can be used to characterize particles. A common method for expressing how much a particle is stretched is the L/W ratio. As seen in Figure 3a, if the L/W ratio is greater than 1, it can be said that the particle is elongated, and if the L/W ratio is very small, it can be said that it is a compact (bulky) or highly spherical particle. Since it was stated in previous studies that the shape of ground mineral particles is best expressed by the Bounding Rectangular Aspect Ratio (*BRAR*) [24] with *DIA* (see Figure 3b), in this study, a new shape concept, called "the degree of sphericity" of particles, is used (Figure 3c) by taking the inverse of *BRAR* (see Equations (2) and (3)) to investigate the relationship between the d_t and roundness values of graphite particles ground at various times in the *VDM*.

$$BRAR = (BRL/BRW) \tag{2}$$

$$1/BRAR = (BRW/BRL) \tag{3}$$

where *BRL* and *BRW* are bounding rectangle length and bounding rectangle width, respectively.

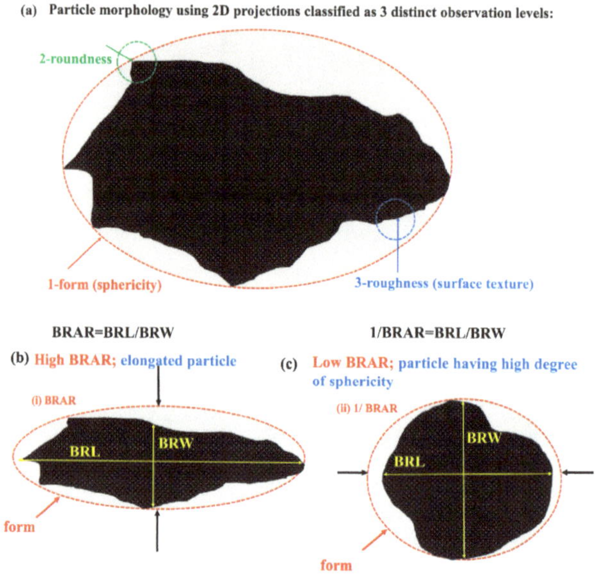

Figure 3. Conceptual diagram of 2D particle morphology redrawn from [25–28]: (**a**) classified as 3 distinct observation levels; (**b**) elongated particle; (**c**) spherical particle.

3. Results

3.1. Effect of Particle Size on d_t Values

Since the tapped bulk density of natural graphite powder used as LIB anode material is generally known to be in the range of 1.0 to 1.2 g/cm^3, the measured d_t values are consistent with the literature. When the d_t values of these size fractions were compared, it was observed that the d_t values in the finer size fraction were generally higher than the d_t values in the coarser size fraction, as shown in Figure 4, even though two fractions that were close to one another were prepared to examine the change in d_t values of the VDM products depending on the grinding time in the same size range. This is well aligned with the previously reported study suggesting that the smaller the average particle size, the lower the d_t [7,10].

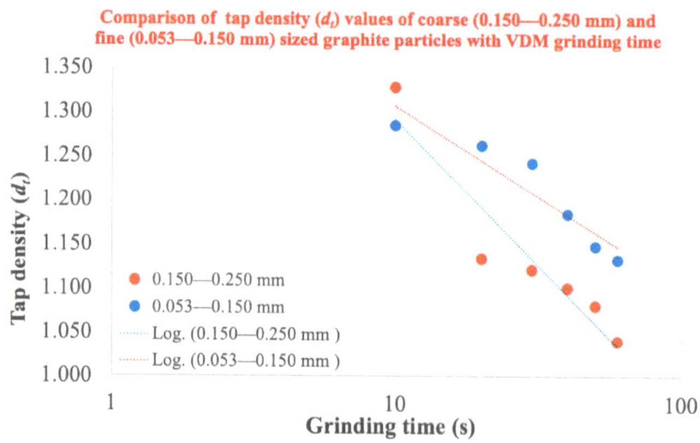

Figure 4. Effect of particle size on changes in d_t of graphite particles ground in the *VDM*, depending on grinding time.

3.2. Empirical Time-Dependent d_t Models

The degree to which a model generalizes to the data on which it is based is called the fitting technique. This allows for the precise prediction of approaching data points with an ideal fitting technique. Techniques that use linear, exponential, and polynomial functions to predict or explain the dependent variable based on the independent variable can capture patterns, relationships, and trends in the data, allowing for more accurate predictions and a deeper understanding of underlying connections [16]. In this section, correlations between the measured average d_t values of graphite particles by a VDM and the milling time were established with various fitting models such as linear, logarithmic, second order polynomial, power, and exponential, and finally compared according to the highest R^2 value, as given in Figure 5.

3.2.1. Linear Fitting Model

In the linear fitted model, the established empirical time-dependent d_t models for the correlations between d_t and grinding time for vibratory disc-milled graphite particles are in the form of "$d_t = -a.t + b$" type equations (where a and b are constants), as shown in Figure 5a,b). In this model, it is seen that the R^2 value of 0.7424 obtained for the coarser size fraction (Figure 5a) is smaller than the R^2 value of 0.9713 obtained for the finer size fraction (Figure 5b).

3.2.2. Logarithmic Fitting Model

The established empirical time-dependent d_t models for the correlations between d_t and grinding time in the logarithmic fitted model take the form of equations of the "$d_t = -a.ln\ (t) + b$" type, where a and b are constants. As seen in Figure 5c,d, the R^2 value obtained for the coarser particle size fraction (0.8944) was found to be higher than the R^2 value obtained for the finer particle size fraction (0.8821). In fact, the d_t data measured for both size fractions were in close agreement with the milling time but not very strong.

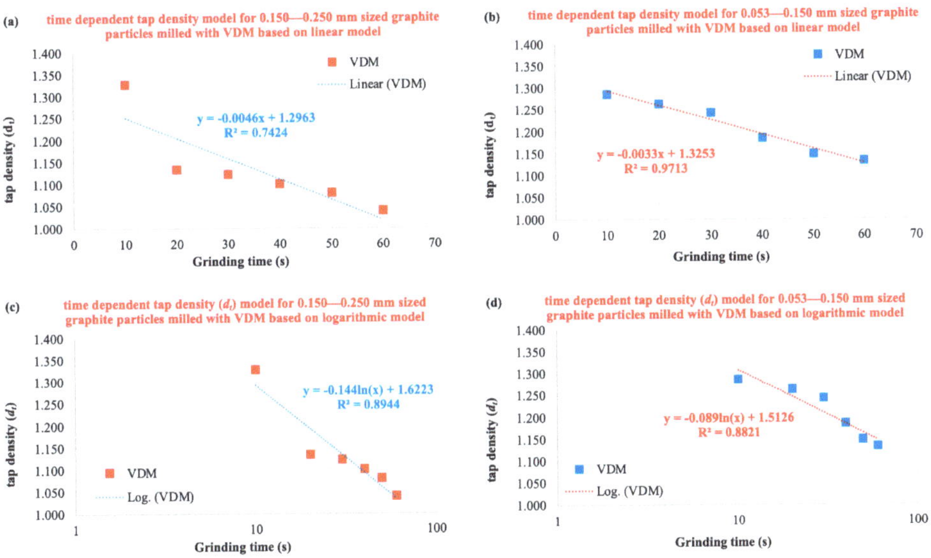

Figure 5. *Cont.*

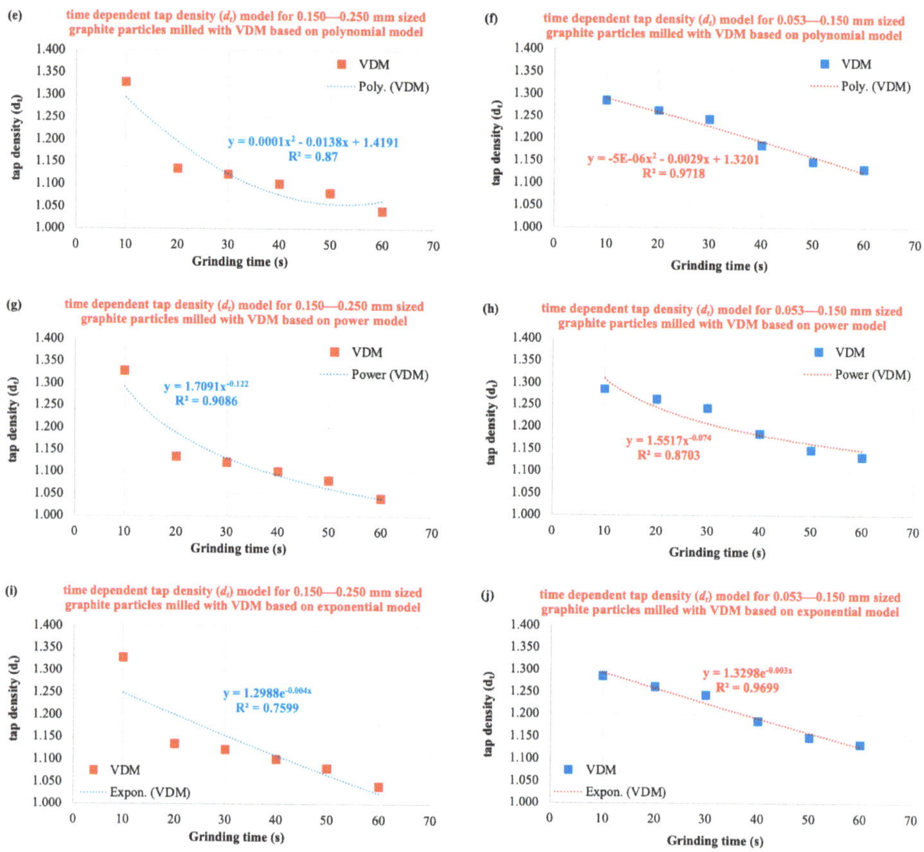

Figure 5. Correlation of d_t values of graphite particles ground in a *VDM* with grinding time according to various fitting models (for 0.150–0.250 mm and 0.053–0.150 mm size fractions). (**a**,**b**) linear; (**c**,**d**) logarithmic; (**e**,**f**) polynomial; (**g**,**h**) power; (**i**,**j**) exponential model.

3.2.3. Polynomial Fitting Model

In the second-order polynomial fitted model, the established empirical time-dependent d_t models for the correlations between d_t and grinding times are in the form of "$d_t = at^2 - bt + c$" type equations (where a, b, and c are constants), as shown in Figure 5e,f. In this model, it is seen that the R^2 value of 0.8700 obtained for the coarser size fraction (Figure 5e) is smaller than the R^2 value of 0.9718 obtained for the finer size fraction (Figure 5f).

3.2.4. Power Fitting Model

In this model, the empirical time-dependent d_t models established for the correlations between d_t and grinding time for graphite by a *VDM* are in the form of equations of the type "$d_t = a.t^{(-b)}$" (where a and b are constants), as shown in Figure 5g,h. When comparing the R^2 values of this model, it was found that the coarser size fraction giving a value of 0.9086 R^2 (Figure 5g) was stronger than the finer size fraction giving a value of 0.8703 R^2 (Figure 5h).

3.2.5. Exponential Fitting Model

In the last fitting model, namely the exponential model, the empirical time-dependent d_t models are in the form of equations of the type "$d_t = a.e^{(-bt)}$" (where a and b are constants), as shown in Figure 5i,j. The correlation between d_t and grinding times in the finer size fraction (0.9699 R^2 in Figure 5j) is considerably stronger than in the coarser fraction (0.7599 R^2 in Figure 5i).

3.2.6. Discussion

The summary of the established equations along with R^2 values is given in Table 1. It was concluded that, while the power model (with a 0.9086 R^2) describes the d_t data for the coarser size fractions better than the alternative fitting models (linear, logarithmic, second-order polynomial, and exponential), the d_t data for the finer size fractions were well represented by the linear, second-order polynomial, and exponential models, considering the highest R^2 value. Particularly, very high R^2 values, such as 0.97, were obtained by the linear, second-order polynomial, and exponential models for this size fraction. Considering both coarse and fine-size fractions, the best fit is the second-order polynomial model.

Table 1. Summary of the results of the established empirical time-dependent d_t models.

Fitting Models	Size Fractions (mm)	Equations	R^2
Linear	0.150–0.250	$d_t = -0.0046t + 1.2963$	0.7424
	0.053–0.150	$d_t = -0.0033t + 1.3253$	0.9713
Logarithmic	0.150–0.250	$d_t = -0.144ln(t) + 1.6223$	0.8944
	0.053–0.150	$d_t = -0.089ln(t) + 1.5126$	0.8821
Second-order polynomial	0.150–0.250	$d_t = 0.0001t^2 - 0.0138t + 1.4191$	0.8700
	0.053–0.150	$d_t = -0.000005t^2 - 0.002933t + 1.320119$	0.9718
Power	0.150–0.250	$d_t = 1.7091t^{(-0.122)}$	0.9086
	0.053–0.150	$d_t = 1.5517t^{(-0.074)}$	0.8703
Exponential	0.150–0.250	$d_t = 1.2988e^{-0.004t}$	0.7599
	0.053–0.150	$d_t = 1.3298e^{-0.003t}$	0.9699

3.3. Validation of the Fitting Models Established: Predicted Versus Experimental d_t Values

To reveal the strength of the relationship shown by the established empirical models in Figure 5, average d_t values were predicted based on the experimental grinding times. While the relationships between predicted and experimental d_t data for both size fractions of ground graphite particles by a VDM based on various fitting models were given in Figure 6, the experimental and predicted values of d_t depending on various grinding times with absolute % errors and relative % errors were summarized in Table 2.

It was concluded that, while the second-order power model describes the d_t data for the coarser size fractions better than the alternative fitting model (linear, logarithmic, second-order polynomial, and exponential), the d_t data for the finer size fractions were well represented by the linear, second-order polynomial, and exponential models, considering the highest R^2 value with the lowest error values.

Considering the highest R^2 values and the lowest error values together, the exponential fitting model was the best model for modeling d_t values according to grinding time.

It should be noted that the derived best-fit relations (e.g., d_t and time) and fitting functions depend on the initial graphite feed size or grinding parameters.

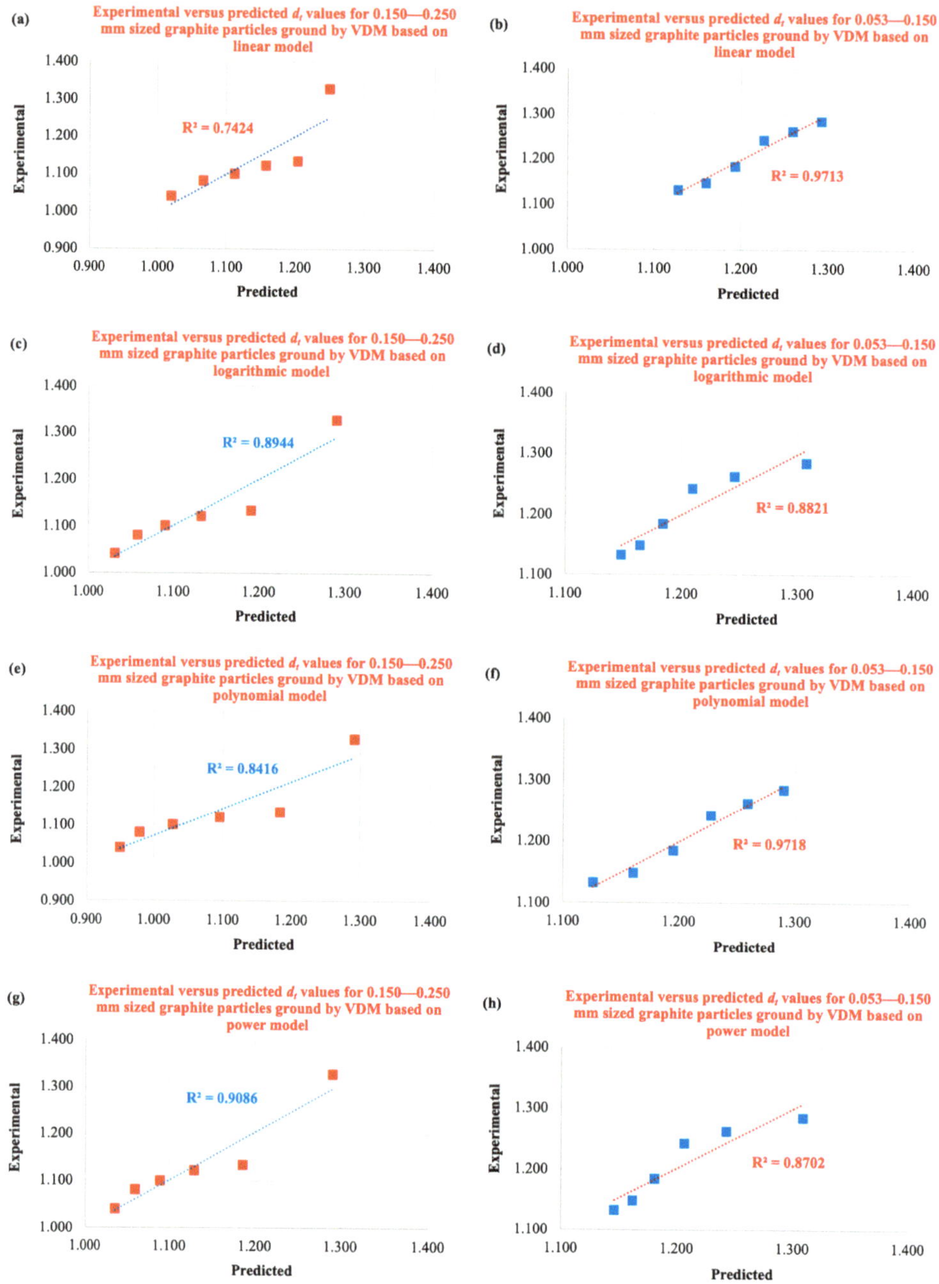

Figure 6. *Cont.*

(i)

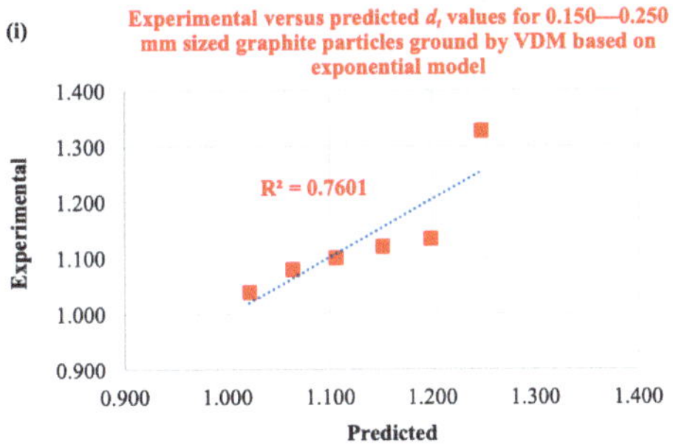

(j)

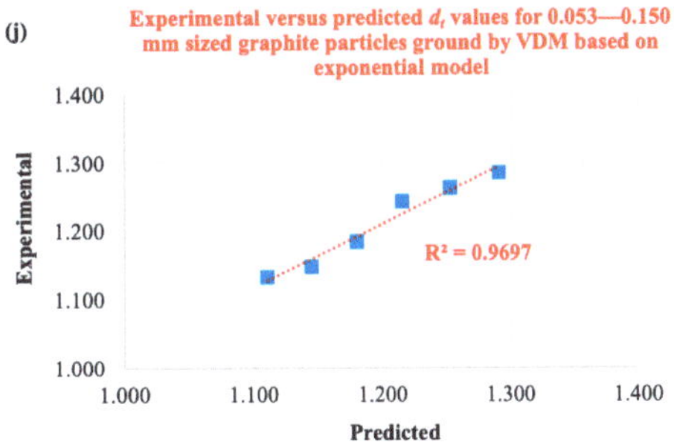

Figure 6. Correlation of predicted versus experimental d_t values of graphite particles ground in a *VDM* according to various fitting models (for 0.150–0.250 mm and 0.053–0.150 mm size fractions), (**a**,**b**) linear; (**c**,**d**) logarithmic; (**e**,**f**) polynomial; (**g**,**h**) power; (**i**,**j**) exponential model.

Table 2. The error values, calculated for the d_t prediction models.

Fitting Model	Size (mm)	Grinding Time	Absolute Error *	Relative Error (%) **
Linear	0.150–0.250	10 s	0.078	5.851
		20 s	−0.070	−6.199
		30 s	−0.037	−3.272
		40 s	−0.012	−1.080
		50 s	0.014	1.324
		60 s	0.020	1.881
	0.053–0.150	10 s	−0.007	−0.577
		20 s	0.003	0.224
		30 s	0.016	1.295
		40 s	−0.009	−0.739
		50 s	−0.012	−1.057
		60 s	0.005	0.465
Logarithmic	0.150–0.250	10 s	0.037	2.807
		20 s	−0.057	−5.019
		30 s	−0.011	−0.974
		40 s	0.009	0.846
		50 s	0.022	2.002
		60 s	0.007	0.687
	0.053–0.150	10 s	−0.023	−1.773
		20 s	0.016	1.280
		30 s	0.033	2.616
		40 s	0.000	0.022
		50 s	−0.016	−1.417
		60 s	−0.016	−1.380
Second-order polynomial	0.150–0.250	10 s	0.037	2.779
		20 s	−0.049	−4.330
		30 s	0.026	2.363
		40 s	0.073	6.662
		50 s	0.102	9.394
		60 s	0.089	8.536
	0.053–0.150	10 s	−0.005	−0.420
		20 s	0.003	0.212
		30 s	0.015	1.188
		40 s	−0.010	−0.865
		50 s	−0.013	−1.115
		60 s	0.006	0.568

Table 2. *Cont.*

Fitting Model	Size (mm)	Grinding Time	Absolute Error *	Relative Error (%) **
Power	0.150–0.250	10 s	0.037	2.822
		20 s	−0.052	−4.575
		30 s	−0.007	−0.628
		40 s	0.011	0.972
		50 s	0.020	1.865
		60 s	0.003	0.262
	0.053–0.150	10 s	−0.024	−1.846
		20 s	0.019	1.502
		30 s	0.036	2.895
		40 s	0.004	0.299
		50 s	−0.014	−1.176
		60 s	−0.014	−1.195
Exponential	0.150–0.250	10 s	0.080	6.034
		20 s	−0.065	−5.727
		30 s	−0.030	−2.704
		40 s	−0.006	−0.577
		50 s	0.017	1.595
		60 s	0.018	1.749
	0.053–0.150	10 s	−0.006	−0.437
		20 s	0.010	0.774
		30 s	0.027	2.177
		40 s	0.005	0.433
		50 s	0.004	0.313
		60 s	0.022	1.927

* Absolute error = (experimental−predicted). ** Relative error = [(experimental−predicted)/(experimental)] × 100.

3.4. Correlation of d_t and Particle Shape

It is known that more suitable anode materials for lithium batteries can be obtained by grinding natural graphite ore in a suitable mill [29]. Thus, in our recent research using the same graphite samples to investigate the change in particle shape with grinding times [16], it has been reported that *VDM* particles (see Figure 7) have more spherical shapes than ball mill particles for both size fractions, and this is due to the greater abrasion force in *VDM* grinding compared to lab-scale ball mill grinding, which has a dominant impact force. Starting from this point, the relationship between the degree of sphericity and d_t of the particles produced with the same ore only in a *VDM* was investigated, and correlations were established between the d_t values and the degree of sphericity values of graphite samples prepared in both size fractions ground by a *VDM*. It was found that the degree of sphericity of the particles was negatively correlated with d_t (see Figure 8). Although it is generally known that there is a positive relationship, there are also conflicting results in the literature [10]. For instance, Li et al. [30] have shown that cylinders (a proxy for elongated particles) can achieve higher ordered packing densities, supporting the hypothesis that alignment during tapping enhances density. In addition, Wang et al. [31] showed that geometrically asymmetric particles composed of poly-superquadric elements have a higher packing density than symmetric particles composed of superquadric elements. In addition, Wouterse et al. [32] presented that the packing density reached its maximum when the particle's aspect ratio was about 1.25. Similarly, Li et al. [33] reported that ellipsoids of certain shapes are more densely packed than spheres. Furthermore, using morphology to estimate bulk density may have low precision due to agglomerated powders, as stated by Pisecky et al. [34] and Ding et al. [35]. Another reason is thought to be the ability of long particles to fit into voids or to interlock to reduce void space. Considering that tapping or vibration applies mechanical energy to the particle system, causing the particles to move and rearrange, the increasingly parallel alignment of the particles that tend to rotate and slide can reduce the gaps between the particles; they can lie closer to each other, and this aligned arrangement can lead to a more compact structure by minimizing the empty space and thus a higher d_t. Furthermore, this situation is attributed to the impurity behaviors in graphite. In other words, using a sample as close to pure as possible from the ore exiting the mine in the experiments was only possible with 72% pure graphite. It is thought that the size, shape, and roughness values of other gangue minerals found with graphite also affect its tap density behavior.

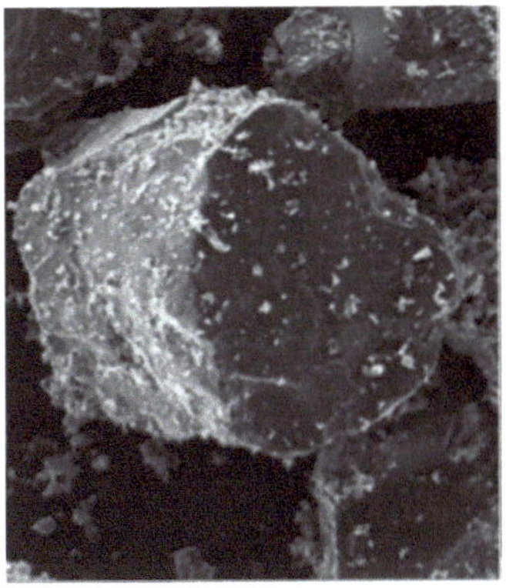

Figure 7. 10 s ground graphite particle in a *VDM* (*SEM* mag: ×500).

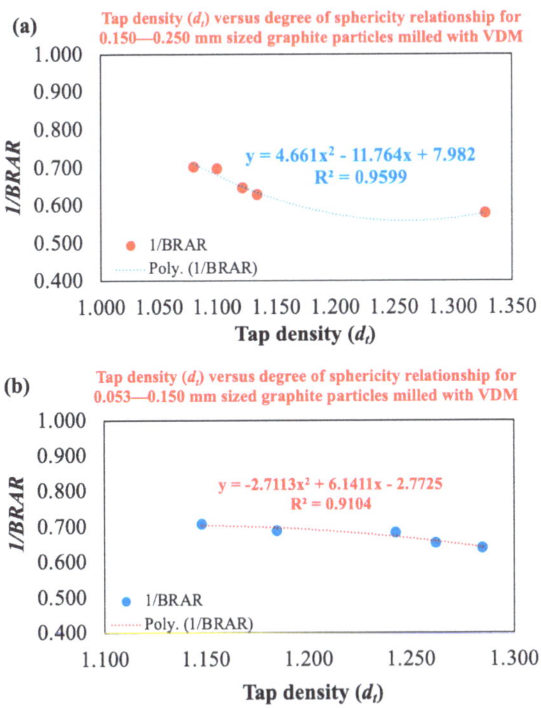

Figure 8. Correlation of d_t and degree of sphericity (*1/BRAR*$_{av.}$), which is based on the *BRAR*$_{av.}$ data of graphite particles by *VDM* with grinding time [16] (**a**) 0.150–0.250 mm and (**b**) 0.053–0.150 mm.

4. Conclusions

Since the d_t of natural graphite particles used as anode material in *LIBs* can generally be increased by optimized grinding to maximize the high energy density, the average d_t values of graphite particles ground by a *VDM* at various grinding times for both size fractions were correlated with the changes in grinding time based on various fitting models.

The best-established correlations are in the form of "$d_t = ae^{-bt}$" type along with the exponential fitting model along with the highest R^2 and the lowest error values.

Considering the high R^2 values obtained, the average d_t values predicted by the established time-dependent mathematical models are in very good agreement with the experimental values.

Since the grinding process is both energy-intensive and cost-intensive, the findings will benefit the graphite industry, which requires appropriate d_t control, powder flow and processing, and packaging of ground natural graphite mineral materials, especially the production of graphite *LIB* anode material with high d_t.

5. Future Recommendation

As of now, the grinding of graphite particles in *LIB* applications has not been optimized; this study will provide a first baseline for the performance that can be expected from different mills as well as other LIB electrode materials. In addition, future studies along these lines could examine the factors that make the exponential model the best fit, aiming to identify the physical characteristics of the material and the process parameters that the fitting constants are serving as proxies for. Moreover, an additional energy efficiency analysis is suggested as a future study, as it would significantly increase the practical value of the research.

Author Contributions: Conceptualization, U.U.; methodology, G.G., B.M., and G.B.; validation, U.U.; investigation, G.G., B.M., and G.B.; resources, F.B.; data curation, U.U.; writing—original draft, U.U.; writing—review and editing, U.U.; visualization, U.U.; supervision, U.U. and F.B. All authors have read and agreed to the published version of the manuscript.

Funding: This research received no external funding.

Data Availability Statement: Dataset available on request from the authors.

Acknowledgments: First, we are grateful to Karabacak Mining Inc. for supplying the graphite samples that were utilized in this investigation. Second, we are grateful to Emrecan Söylemez of the Faculty of Mechanical Engineering at Istanbul Technical University for permitting us to utilize the tap density instrument at their department. Thirdly, the laboratory facilities offered by Istanbul Technical University's Mineral Processing Engineering Department and Sivas Cumhuriyet University's Mining Engineering Department are also valued. Fourth, we are grateful to Halil İbrahim Çetintas from the Advanced Technology Application and Research Center and İlhan Danacı from the Department of Aeronautical Engineering at Sivas Science and Technology University for their XRD and SEM studies of the samples. Lastly, the authors would like to express their gratitude to Ezgi Onur Şahin of the Max-Planck-Institut für Kohlenforschung in Germany for her invaluable assistance in interpreting the graphite sample's mineralogical composition based on the XRD results.

Conflicts of Interest: The authors declare no conflicts of interest.

Abbreviations

1/BRAR	Inverse of bounding rectangular aspect ratio; degree of particle sphericity
$BRAR_{av.}$	Average value of bounding rectangular aspect ratio
BRL	Bounding rectangle length

BRW	Bounding rectangle width
DIA	Dynamic Image Analysis
d_t	Tap density
EDS	X-ray spectroscopy
HEVs	Hybrid electric vehicles
LIB	Lithium-ion battery
m	Mass of the powder (gr)
R^2	Coefficient of determination
SEM	Scanning Electron Microscopy
V	Volume of the powder (g/cm^3)
VDM	Vibrating Disc Mill
XRD	X-Ray Diffraction

References

1. Future Battery Industries CRC. Natural Graphite Purification Report. AusIndustry Cooperation Research Centres Program. September 2022. Available online: https://fbicrc.com.au/wp-content/uploads/2023/11/FBICRC_Graphite-Purification-Report-VFINAL-Sept22.pdf (accessed on 27 February 2003).
2. Schalkwijk, W.A.; Scrosati, B. *Advances in Lithium-Ion Batteries*; Kluwer Academic: New York, NY, USA, 2002.
3. Qi, Y.; Guo, H.; Hector, L.G., Jr.; Timmons, A. Threefold Increase in the Young's Modulus of Graphite Negative Electrode during Lithium Intercalation. *J. Electrochem. Soc.* **2010**, *157*, A558–A566. [CrossRef]
4. Liu, P. *Improving the Tapped Density of the Cathode Material to Make a Lithium-Ion Battery Hold More Energy*; Application Notes; Application Research Lab, Bettersize Instruments Ltd.: Liaoning, China, 2021. Available online: https://www.bettersizeinstruments.com/learn/knowledge-center/improving-the-tapped-density-of-the-cathode-material-to-make-a-lithium-ion-battery-hold-more-energy/ (accessed on 27 February 2003).
5. Lee, G.-H.; Yi, H.; Kim, Y.-J.; Lee, J.B.; An, J.-C.; Park, S.-M.; Oh, K.; Yoon, S.-H.; Park, J.-I. A Shortened Process of Artificial Graphite Manufacturing for Anode Materials in Lithium-Ion Batteries. *Processes* **2024**, *12*, 2709. [CrossRef]
6. IRENA. *Geopolitics of the Energy Transition: Critical Materials*; International Renewable Energy Agency: Abu Dhabi, United Arab Emirates, 2023. Available online: https://www.irena.org/-/media/Files/IRENA/Agency/Publication/2023/Jul/IRENA_Geopolitics_energy_transition_critical_materials_2023.pdf (accessed on 27 February 2003).
7. Ying, J.; Wan, C.; Jiang, C.; Li, Y. Preparation and characterization of high-density spherical $LiNi_{0.8}Co_{0.2}O_2$ cathode material for lithium secondary batteries. *J. Power Sources* **2001**, *99*, 78–84. [CrossRef]
8. Yang, S.; Wang, X.; Yang, X.; Liu, Z.; Wei, Q.; Shu, H. High Tap Density Spherical $Li[Ni_{0.5}Mn_{0.3}Co_{0.2}]O_2$ Cathode Material Synthesized via Continuous Hydroxide Coprecipitation Method for Advanced Lithium-Ion Batteries. *Int. J. Electrochem.* **2012**, *9*, 323560. [CrossRef]
9. Zhang, J.-p.; Wang, D.-k.; Zhang, L.-h.; Liu, H.-y.; Liu, Z.-b.; Xing, T.; Ma, Z.-k.; Chen, X.-h.; Song, H.-h. A wet granulation method to prepare graphite particles with a high tap density for high volumetric energy density lithium-ion storage. *New Carbon Mater.* **2022**, *37*, 402–410. [CrossRef]
10. Mazlan, M.R.; Jamadon, N.H.; Sulong, A.B.; Aripin, M.A.; Jamhari, F.I.; Yusof, F. Comparative Study of Particle Size and Shape Effects on Powder Packing Densities. *J. Kejuruter.* **2024**, *36*, 2563–2570. [CrossRef]
11. Bettersize. BeDensi T Pro Series. 2025. Available online: https://www.bettersizeinstruments.com/products/bedensi-t-series-tapped-density-meter/ (accessed on 27 February 2003).
12. Azom.com. Improving Lithium-Ion Batteries Through Measuring Tapped Density. 2021. Available online: https://www.azom.com/article.aspx?ArticleID=21056 (accessed on 27 February 2003).
13. Jeswiet, J.; Szekeres, A. Energy Consumption in Mining Comminution. *Procedia CIRP* **2016**, *48*, 140–145. [CrossRef]
14. Seino, K.; Golman, B.; Shinohara, K.; Ohzeki, K. Variation of packing structure of cast film with preparation conditions and particle properties. *TANSO* **2005**, *216*, 2–7. [CrossRef]
15. Qiu, X. *Investigating the Particle Size and Shape Influences on Anode Energy Density of Lithium-Ion Batteries*; Bettersize Application Note; Application Research Lab, Bettersize Instruments Ltd.: Liaoning, China, 2024. Available online: https://www.bettersizeinstruments.com/uploads/file/investigating-the-particle-size-and-shape-influences-on-anode-energy-density-of-lithium-ion-batteries.pdf (accessed on 27 February 2003).
16. Ulusoy, U.; Burat, F.; Bayar, G.; Mojtahedi, B.; Güven, G. Modeling the change of the sphericity feature of graphite particles ground in a ball and vibrating disc mill with grinding time. *J. Energy Storage* **2024**, *97*, 112814. [CrossRef]
17. Andrić, L.; Trumić, M.; Trumić, M.; Nikolić, V. Micronization of zeolite in vibration mill. *RSD* **2018**, *11*, 63–71. [CrossRef]
18. *ISO 3953:2011*; Metallic Powders—Determination of Tap Density. Technical Committee ISO/TC 119: Stockholm, Sweden, 2011.

19. Ulusoy, U. Dynamic image analysis of differently milled talc particles and comparison by various methods. *Part. Sci. Technol.* **2018**, *36*, 332–339. [CrossRef]
20. Ulusoy, U. Quantifying of particle shape differences of differently milled barite using a novel technique: Dynamic image analysis. *Materialia* **2019**, *8*, 100434. [CrossRef]
21. Czajkowska, M.; Sznitowska, M.; Kleinebudde, P. Determination of coating thickness of minitablets and pellets by dynamic image analysis. *Int. J. Pharm.* **2015**, *495*, 347–353. [CrossRef] [PubMed]
22. Weger, R.J.; Eberli, G.P.; Baechle, G.T.; Massaferro, J.L.; Sun, Y.-F. Quantitative characterization of carbonate pore systems by digital image Analysis. *AAPG Bull.* **1998**, *93*, 1297–1317. [CrossRef]
23. Ma, G.; Bu, X.; Ulusoy, U.; Xie, G. Effect of particle shape on bubble-particle attachment behavior: Roles of surfaces, edges and vertexes. *J. Clean. Prod.* **2023**, *429*, 139606. [CrossRef]
24. Particle Insight Manual. Micromeritics Instrument Corp.: Norcross, GA, USA, 2013.
25. Mitchell, J.K.; Soga, K. *Fundamentals of Soil Behaviour*, 3rd ed.; John Wiley & Sons: Hoboken, NJ, USA, 2005.
26. Barrett, P.J. The shape of rock particles, a critical review. *Sedimentology* **1980**, *27*, 291–303. [CrossRef]
27. Blott, S.J.; Pye, K. Particle shape: A review and new methods of characterization and classification. *Sedimentology* **2008**, *55*, 31–63. [CrossRef]
28. Ulusoy, U. A review of particle shape effects on material properties for various engineering applications: From macro to nanoscale. *Minerals* **2023**, *13*, 91. [CrossRef]
29. Disma, F.; Aymard, L.; Dupont, L.; Tarascon, J.-M. Effect of Mechanical Grinding on the Lithium Intercalation Process in Graphites and Soft Carbons. *J. Electrochem. Soc.* **1996**, *143*, 3959–3972. [CrossRef]
30. Li, S.; Zhao, J.; Lu, P.; Xie, Y. Maximum packing densities of basic 3D objects. *Chin. Sci. Bull.* **2010**, *55*, 114–119. [CrossRef]
31. Wang, S.; Marmysh, D.; Ji, S. Construction of irregular particles with superquadric equation in DEM. *Theor. Appl. Mech. Lett.* **2020**, *10*, 68–73. [CrossRef]
32. Wouterse, A.; Williams, S.R.; Philipse, A.P. Effect of particle shape on the density and microstructure of random packings. *J. Phys. Condens. Matter.* **2007**, *19*, 406215. [CrossRef] [PubMed]
33. Li, C.X.; Gan, J.Q.; Pinson, D.; Yu, A.B.; Zhou, Z.Y. Dynamic analysis of poured packing process of ellipsoidal particles. *Powder Technol.* **2021**, *385*, 444–454. [CrossRef]
34. Pisecky, J. *Handbook of Milk Powder Manufacture*; GEA Process Engineering A/S: Soeborg, Denmark, 2012.
35. Ding, H.; Li, B.; Boiarkina, I.; Wilson, D.I.; Yu, W.; Young, B.R. Effects of Morphology on the Bulk Density of Instant Whole Milk Powder. *Foods* **2020**, *9*, 1024. [CrossRef]

minerals

MDPI

Article

Experimental Study on the Influence of Rotational Speed on Grinding Efficiency for the Vertical Stirred Mill

Biliang Tang [1,2,3], Bo Cheng [2], Xianzhou Song [1,2,3], Haonan Ji [4], Yijiang Li [4] and Zhaohua Wang [1,3,4,*]

1 CITIC Heavy Industries Co., Ltd., Luoyang 471039, China; tangbl@citic-hic.com.cn (B.T.); songxz@citic-hic.com.cn (X.S.)
2 National Key Laboratory of Intelligent Mining Heavy Equipment, Luoyang 471039, China; chengbo@citic-hic.com.cn
3 Luoyang Mining Machinery Engineering Design Institute Co., Ltd., Luoyang 471039, China
4 School of Mechanical Engineering, Taiyuan University of Science and Technology, Taiyuan 030024, China; s202312210094@stu.tyust.edu.cn (H.J.); liyijiang_cumt@163.com (Y.L.)
* Correspondence: wangzhaohua@tyust.edu.cn; Tel.: +86-15533545658

Abstract: The rotational speed of the agitator is one of the important parameters that affect the grinding efficiency of the vertical stirred mill. Increasing the speed will improve the grinding effect, but it will increase energy consumption, and determining a reasonable speed setting is a system issue. The effects of different speeds on energy consumption, product particle size, and grinding efficiency were analyzed in this study. An experimental vertical stirred mill was used to grind iron ore, and five different speed parameters from 175 rpm to 350 rpm were set as variables. It was found that increasing the rotational speed will increase the grinding effect, but it will trigger more energy consumption. A new evaluation index to comprehensively reflect the grinding efficiency of the mill, which was defined as the ability of a mill to grind the same product per unit of time and energy consumption, was proposed. The grinding efficiency was calculated when the particle size of iron ore powder decreased to −45, −38, and −28 μm at different speeds. It can be seen that the growth rate of energy consumption is faster than that of the percentage of particle size, which leads to a continuous decrease in grinding efficiency with the increase in rotational speed. If high processing capacity is pursued within a certain period of time, high speed can be chosen, but it will result in energy loss. On the contrary, the low speed can be chosen, if considering grinding economy.

Keywords: vertical stirred mill; rotational speed; grinding efficiency; energy consumption; particle size

1. Introduction

Milling is one of the most energy intensive processes in industry, contributing 2% of all global energy usage [1]. As the reserves of many nonrenewable mineral resources around the world are decreasing year by year, and the resource occurrence grade is low, the particle size is fine, and the composition is complex, a large amount of poor fine impurities and difficult-to-process ores are problematic to comprehensively utilize due to outdated grinding equipment, high energy consumption, and low efficiency [2]. A vertical stirred mill is a type of energy-saving ultrafine milling equipment that mainly works by friction grinding. Compared to traditional horizontal ball mills, it can reduce production energy consumption for coarse grinding by up to 30% [3] and has the advantages of low energy consumption, low noise, and a small footprint. It is widely used in mineral processing [4].

During the operation of the vertical stirred mill, the motor drives the agitator to rotate, which in turn drives the circular reciprocating motion of the grinding media and slurry inside the cylinder. Under the continuous squeezing, collision, and shearing of the grinding spheres, the micro-sized ore particles in the slurry are refined [5–7]. There is a multiphase material coupling effect between the agitator, slurry, and grinding spheres inside the cylinder, and the motion law is very complex. Different operating parameters, structural parameters, and process parameters will result in different grinding efficiencies [8,9]. The local grinding media concentration, velocity profiles, grinding media collisions, and stress energies were compared for varied total grinding media fillings and stirrer speeds by Fragnière et al. [10]. Rhymer et al. explored the fundamental dynamics of vertical stirred mills when using multiple sizes of grinding media by employing the discrete element method (DEM) [11], and the grinding performance was evaluated from five aspects: media segregation, media velocity, media force, contact energy, and power draw. A HIGmill, operated on a copper regrind circuit, was sampled under different operating conditions, i.e., tip speeds, solids contents, and media filling, and the influence on grinding efficiency was studied by Altun et al. [12]. The influence of different operating parameters on the temperature of a stirred mill was studied by Guner et al. [13], and a power number correlation was established to calculate the power under any milling condition, which determines the heat generation rate. The effects of agitator shape, including tower mill and pin mill, and the shape of the grinding media on grinding efficiency were studied by Sinnott et al. [14,15]. Therefore, reasonable parameter settings can improve the grinding efficiency of a vertical stirred mill.

The rotational speed of the agitator is one of the important parameters affecting the grinding efficiency. Several researchers found that the increase in speed resulted in an increase in fineness and specific surface area of the product [16,17]. Oliveira et al. [18] proposed that the speed had a very significant effect on predicted apparent breakage rates, increasing approximately in proportion to the mill's specific power. Such variation was mainly due to the increase in collision frequency with speed. Increase in speed also resulted in an increase in fineness of the apparent breakage function. Fadhel et al. [19] emphasized that increasing the speed of the stirrer would increase the collision force of the grinding spheres and also increase the number of collisions. The grinding efficiency is the highest when the collision energy of the grinding spheres reaches just enough to overcome the particle crushing energy. Mankosa et al. [16,20] proposed that when the rotational speed reaches 7 m/s, vortices will form in the chamber of gravity-induced mills, which will reduce the grinding efficiency. Low speed can actually reduce the influence of vortices and improve grinding efficiency [21]. Rhymer et al. [1] mentioned that higher rotational speeds can improve grinding efficiency, but this comes at the cost of increased power consumption, and the two are conflicting. It is necessary to find a reasonable rotational speed to balance the effectiveness and efficiency of the mill. From this, it can be seen that the rotational speed has a significant impact on the grinding efficiency. Increasing the speed will improve the grinding effect, but it will increase energy consumption. There is still controversy over whether to choose high or low speeds for the mill.

Simulation and experimentation are the main means of studying the operating parameters of a vertical stirred mill [22–24]. The early application of DEM was used to simulate the motion law and wear behavior of grinding spheres in dry stirred mills [25,26]. Subsequently, computational fluid dynamics (CFD) was developed to simulate slurry, which belongs to the fluid phase, and the DEM–CFD method was widely used in wet stirred mills [27,28]. In order to improve the computational efficiency of coupling simulation between slurry and grinding media, the smoothed particle hydrodynamics (SPH) method was developed to model the slurry [29]. Then, the DEM–SPH method was widely used in wet vertical stirred mills [30,31]. Furthermore, the coupled particle finite element method (PFEM), finite element method (FEM), and DEM models were used to simulate the mechanical behavior of a stirred mill [32–34]. Although these simulation methods can reflect grinding behavior through power, force, energy, wear, etc., they cannot truly provide information on product particle size and processing capacity. On the contrary, experiments can provide information such as grinding time and product particle size, which is still the preferred research method for studying grinding efficiency [35–37].

In this study, a vertical spiral stirred mill was taken as the research object, and the effects of different speeds on energy consumption and product particle size were experimentally studied. The purpose of this study is to obtain the influence of speed on energy consumption and product particle size and to establish an evaluation index to comprehensively reflect the grinding efficiency at different speeds, providing a basis for selecting the speed of the mill. The sections are organized as follows. The experimental setup and parameters are introduced in Section 2. The grinding results are provided from two aspects, energy consumption and product size, in Section 3. A new evaluation index of grinding efficiency is established, and the grinding efficiency at different rotational speeds is calculated in Section 4. Finally, some important conclusions are presented in Section 5.

2. Experiment Setup

2.1. Experimental Equipment

In this study, the CSM-2.2 vertical stirred mill (CITIC Heavy Industries Co., Ltd., Luoyang, China), shown in Figure 1, is selected as the experimental equipment, and the grinding effect at different speeds could be determined through batch grinding experiments. The operating power of the motor is 2.2 kW, and it is connected to the spiral agitator through a coupling. A torque and speed tester is installed on the upper end of the coupling to obtain the real-time operating speed and torque of the spiral agitator.

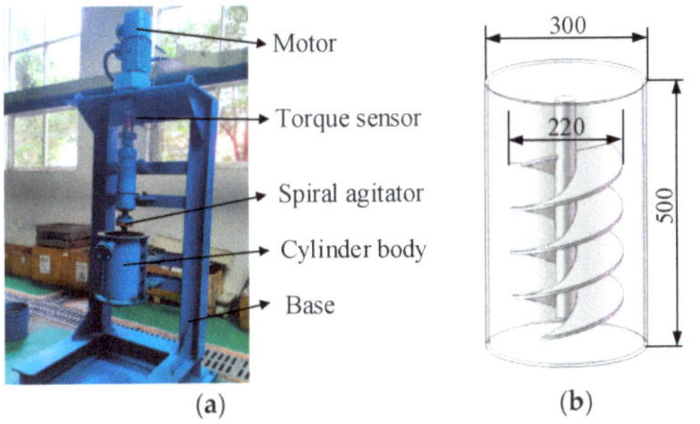

Figure 1. Experimental prototype and internal structure of the cylinder: (**a**) experimental prototype; (**b**) schematic diagram of the internal structure of the cylinder.

The mill specifications are shown in Table 1. The agitator speed is 173–350 (r/min); the outer diameter and height of the cylinder are 300 mm and 500 mm, respectively. Its effective volume is 36 L, and the material is 316 stainless steel. The agitator is manufactured by Q355B, with a maximum outer diameter of 220 mm and a pitch of 175 mm. The feed slurry to the chamber is pumped via the top. After being subjected to the combined action of the agitator and the grinding media inside the cylinder for a period of time, samples can be taken from the overflow port at the upper end.

Table 1. Mill specifications.

Parameter	Value
Installed mill motor power (kW)	2.2
Rotational range of the spiral agitator (r/min)	173–350
Outer diameter of the cylinder (mm)	300
Height of the cylinder (mm)	500
Outer diameter of the spiral agitator (mm)	220
Net volume (m^3)	0.036

2.2. Feed Particle Size

In this study, Donganshan hematite ore is used for grinding experiments. The samples were collected from Liaoning Province, China. The mineral structure is mainly manifested as the automorphic crystal structure of hematite, the semi automorphic crystal structure of magnetite, and the alternating structure between two or more minerals. The chemical composition was detailed separately in Table 2. TFe refers to total Fe, which includes iron elements in various compounds such as iron carbonate, iron oxide, etc. Its content was measured using the photometric titration method, according to the Chinese National Standard GB/T 6730.73 [38]. The FeO, other oxides, and elements were measured using the wet analysis methods according to standard GB/T 6730.71 [39]. The TFe grade in the ore is 33.42%, the FeO content is 4.81%, the SiO_2 content is 46.92%, and the main vein mineral is quartz. The density is 3400 kg/m^3, and the Brinell hardness coefficient is 15.

Table 2. Chemical composition of the samples (wt%).

Number	TFe	FeO	SiO_2	MgO	Al_2O_3	CaO	P	S
Content	33.42	4.81	46.92	0.49	0.86	0.56	0.043	0.026

Before the experiment, 11.2 kg of iron ore powder and 6.72 L of water are mixed to obtain a slurry with a mass concentration of 62.5% and a density of 1957.4 kg/m^3. The particle size distribution of the feed sample measured by a laser particle size analyzer (Malvern Panalytical Mastersizer, Malvern, England) is shown in Figure 2.

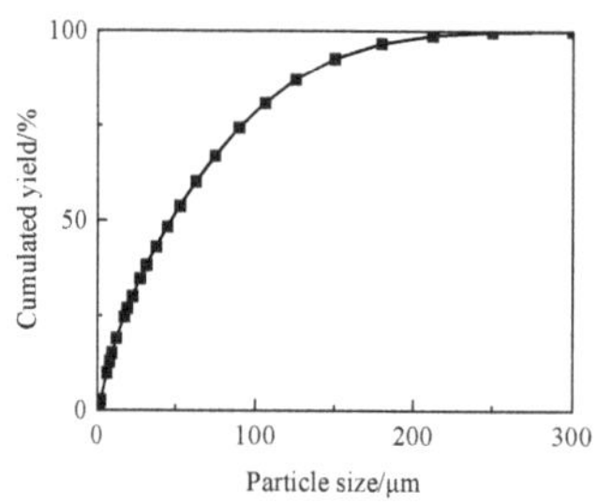

Figure 2. Particle size distribution of feed samples.

2.3. *Experimental Parameters*

Grinding experiments were performed in a wet stirred mill with varying rotational speeds. The steel spheres with a diameter of 8 mm were added inside the cylinder, with a total filling amount of 49 kg, accounting for 48.5% of the cylinder volume. Five agitator speeds were set, as shown in Table 3, including 175, 215, 260, 300, and 350 r/min, corresponding to test numbers A, B, C, D, and E, respectively. After starting the motor, the torque of the agitator was recorded every 10 min, and samples were obtained from the overflow port. A laser particle size analyzer was used to measure the content of −45, −38, and −28 μm in the samples.

Table 3. Agitator speed and test number.

Number	A	B	C	D	E
Speed (r/min)	175	215	260	300	350

3. Experiment Results

3.1. *Energy Consumption*

Each experiment lasted for 30 min, and the torques of the agitator were measured. Figure 3 shows the torque variation pattern of Group A. It can be seen that the torque fluctuates greatly in the initial stage (t = 0–1 min). After running for 1 min, the torque fluctuates within a certain range, indicating that the mill is operating in a stable phase. The time points with significant fluctuations in torque values are excluded, and the average torque during the stable phase is selected as the torque for that time period, corresponding to the first 10 min, second 10 min, and third 10 min, respectively. It can be seen that in the same set of experiments, there is only a small difference in the torque values of the three sampled values, indicating that the vertical stirred mill has reached a stable stage during sampling.

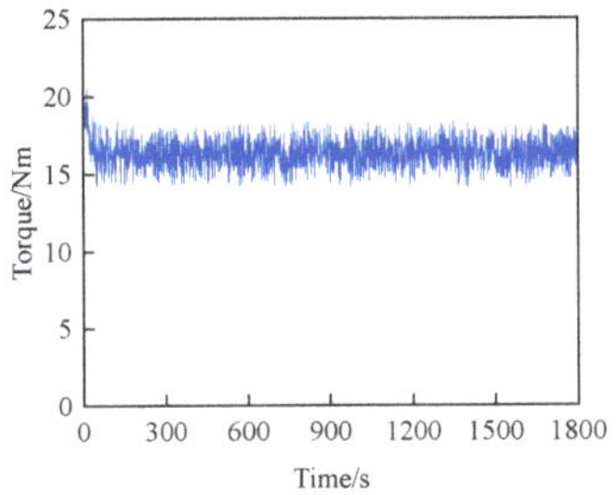

Figure 3. Experiment torque of Group A during 30 min.

Comparing the average torque of each speed with the three sampled values, the trend of its variation is shown in Figure 4. It can be seen that as the speed increases, the torque value of the agitator increases continuously.

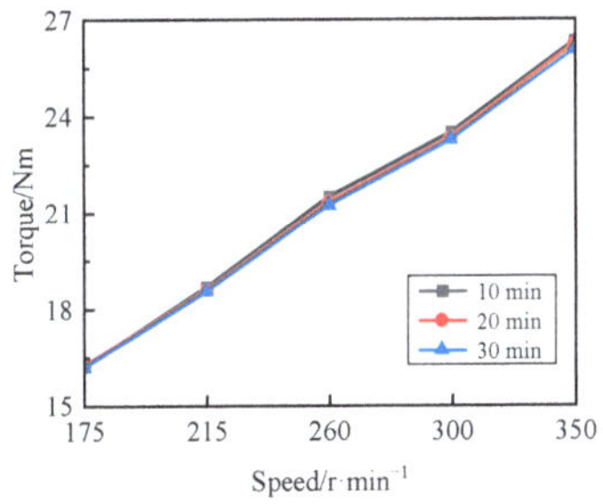

Figure 4. Relationship between the torque and speed of the agitator.

The energy consumption of the mill is not only related to torque, but also to operating speed. The energy consumption E can be calculated by multiplying power P and time t. The relationship between power P and torque T is shown in Equation (1) [40], and the results are shown in Figure 5.

$$P = \frac{T \times n}{9550} \tag{1}$$

where n represents rotational speed.

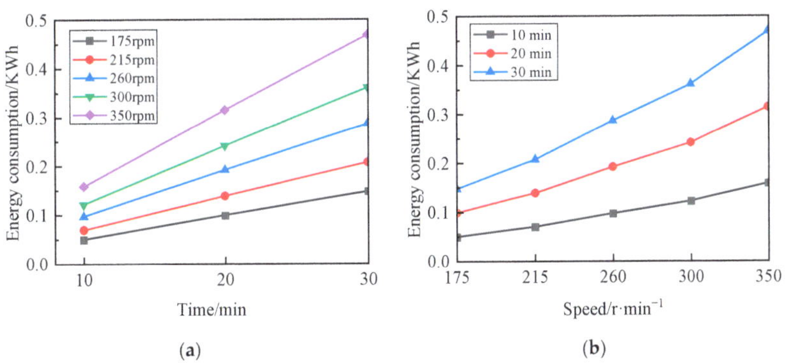

Figure 5. The energy consumption of the mill: (**a**) the variation law of energy consumption over time; (**b**) the variation law of energy consumption over the speed of the agitator.

It can be seen from Figure 5a that at the same rotational speed, energy consumption increases linearly with time. Figure 5b shows that energy consumption increases nonlinearly with speed, and the rate of energy consumption increase is greater with the increase in speed, which can be obtained from the slope in the graph.

3.2. Particle Size of the Product

The particle size of the product can reflect the actual grinding effect of the mill under a set of given experimental parameters. During the experiment, samples are taken from the overflow port every 10 min, and the particle percentage of −45, −38, and −28 μm in the samples is measured using a laser particle size analyzer. The particle size at P_{80} is analyzed.

The experimental results are shown in Table 4. Among them, the number A10 represents the sampling data of Group A at 10 min, and other numbers are used in sequence. The variation law of product percentage with speed is shown in Figure 6.

Table 4. Percentage of particle size at different times and speeds.

Number	−45 μm Particle Size Percentage (%)	−38 μm Particle Size Percentage (%)	−28 μm Particle Size Percentage (%)	Particle Size of P_{80} (μm)
A10	71.51	64.42	52.44	54.52
A20	83.63	76.83	63.53	41.07
A30	92.45	87.52	75.44	31.15
B10	77.32	70.02	57.24	47.83
B20	90.42	84.95	72.50	33.43
B30	96.34	92.90	82.65	26.21
C10	81.26	74.17	60.84	43.58
C20	93.24	88.80	77.63	29.63
C30	99.35	97.45	88.99	22.59
D10	83.32	76.61	63.50	41.33
D20	95.05	91.29	80.78	27.44
D30	99.60	97.96	90.41	21.44
E10	87.63	81.09	67.68	37.09
E20	96.89	93.39	83.32	25.75
E30	99.96	99.52	95.03	19.41

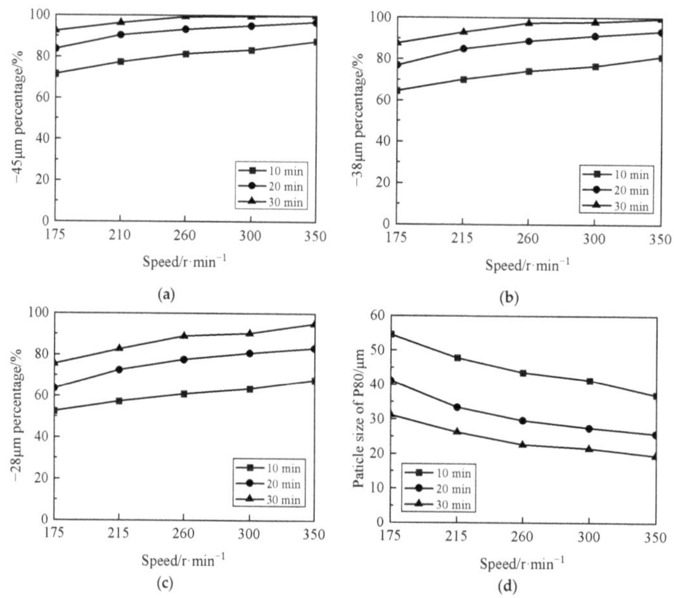

Figure 6. The variation law of particle size over various times and speeds: (**a**) −45 μm; (**b**) −38 μm (**c**) −28 μm; (**d**) particle size of P_{80}.

From Figure 6a–c, it can be seen that at the same rotational speed, as the grinding time increases, the percentage of the same product particle size in the screened material continues to increase. For example, in group A with a speed of 175 r/min, the product percentage of −45 µm after grinding for 10 min is 71.51%. After grinding for 20 min, it increases to 83.63%, and then it increases to 92.45% at 30 min. The variation law of the particle size of other products with grinding time at the same speed follows this law.

Further analysis shows that under the same grinding time, as the agitator speed increases, the percentage of the same product particle size in the screened material also increases continuously. For example, in Group A, with a speed of 175 r/min, the product percentage of −38 µm in Group A10 is 64.42% after grinding for 10 min, and the speed increased sequentially; the product percentages of −38 µm in groups B10, C10, D10, and E10 are 70.02%, 74.17%, 76.61%, and 81.09%, respectively.

The size of P_{80} is calculated to reflect the particle size of the product with a percentage of 80% after grinding for a certain period of time under different parameters. As shown in Figure 6d, at the same rotational speed, the particle size of P_{80} continuously decreases with the increase in grinding time. For example, in Group A with a speed of 175 r/min, the particle sizes of P_{80} after grinding for 10, 20, and 30 min are 54.52, 41.07, and 31.15 µm, respectively. On the other hand, under the same grinding time, as the agitator speed increases, the particle size of P_{80} continuously decreases. For example, in Groups A30, B30, C30, D30, and E30, the particle sizes of P80 are 31.15, 26.21, 22.59, 21.44, and 19.41 um, respectively, after 30 min. From the above analysis, it can be seen that at the same rotational speed, the grinding effect increases with the increase in grinding time. Under the same grinding time, the grinding effect increases with the increase in rotation speed.

After grinding at different speeds for a certain period of time, the energy consumption of the particle size to reach the same percentage is measured. The energy consumption when the percentage of −45, −38, and −28 µm reaches 90%, 80%, and 80% is shown in Table 5. The variation law of energy consumption with speed is shown in Figure 7.

Table 5. Energy consumption when percentage of −45, −38, and −28 µm reaches 90%, 80%, and 80%.

Number	−45 µm (90%)	−38 µm (80%)	−28 µm (80%)
A	0.1263	0.1020	0.1821
B	0.1331	0.1079	0.1837
C	0.1447	0.1174	0.2041
D	0.1593	0.1341	0.2305
E	0.1742	0.1441	0.2579

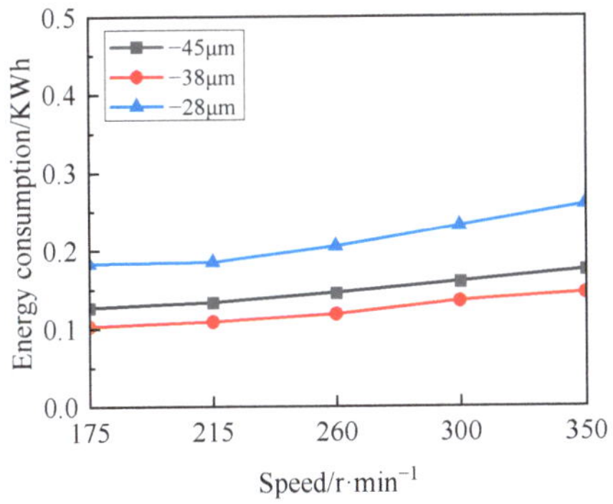

Figure 7. The variation law of energy consumption with rotational speed when the particle size reaches the same percentage.

It can be seen that the energy consumption increases with the increase in rotational speed when the particle size of the same product reaches the same proportion. For example, when the particle percentage of −45 μm reaches 90%, the energy consumption in Group A with a speed of 175 r/min is 0.1263 kWh. As the speed increases, the energy consumption of Groups B, C, D, and E increases sequentially to 0.1331, 0.1447, 0.1593, and 0.1742 kWh, respectively. From this, it can be seen that increasing the rotational speed will increase the grinding effect, but it will trigger more energy consumption. This results is consistent with the results obtained by Rhymer et al. [1], who mentioned that higher rotational speeds will lead to increased effectiveness, but it also comes at a cost of increased power draw, and a compromise would need to be made between effectiveness and efficiency. Some similar conclusions can also be found in References [17,18]. Therefore, the evaluation of grinding efficiency requires the comprehensive consideration of grinding time, energy consumption, and product particle size.

4. Comprehensive Evaluation of Grinding Efficiency

4.1. Evaluation Index

A new evaluation index to comprehensively reflect the grinding efficiency of the mill is proposed in this study. Firstly, for the same sample, the grinding efficiency is influenced by the combined effects of grinding time, product particle size, and energy consumption. The grinding efficiency η can be defined as the ability of a mill to grind the same product per unit of time and energy consumption, which can be calculated by Equation (2). η is a dimensionless evaluation index, and the value is a relative quantity rather than an absolute quantity. For the same group of grinding experiments, a larger value indicates higher grinding efficiency. If the value is normalized, the highest grinding efficiency is 1.

$$\eta_{vt} = \frac{\omega_{vt}}{\frac{E_{vt}}{\overline{E_t}}} \tag{2}$$

where v is the rotational speed, η_{vt} represents the grinding efficiency at a certain speed v and grinding time t, ω_{vt} is the product percentage, and E_{vt} is the energy consumption. $\overline{E_t}$ is the average energy consumption.

4.2. Result Analysis

The energy consumption and the particle percentage of −45, −38, and −28 μm have been shown in Figures 5 and 6. By substituting the above data into the Equation (2), the grinding efficiency can be calculated, and the results are shown in Tables 6–8.

Table 6. Grinding efficiency of −45 μm at different speeds.

Number	A	B	C	D	E
10 min	1.44	1.10	0.83	0.68	0.55
20 min	1.67	1.29	0.96	0.78	0.61
30 min	1.85	1.37	1.02	0.81	0.63

Table 7. Grinding efficiency of −38 μm at different speeds.

Number	A	B	C	D	E
10 min	1.30	1.00	0.76	0.62	0.51
20 min	1.54	1.21	0.91	0.74	0.58
30 min	1.75	1.32	1.00	0.80	0.62

Table 8. Grinding efficiency of −28 μm at different speeds.

Number	A	B	C	D	E
10 min	1.06	0.82	0.62	0.52	0.42
20 min	1.27	1.03	0.80	0.66	0.52
30 min	1.51	1.17	0.91	0.74	0.60

By plotting the data in the table into Figure 8, the variation of grinding efficiency with grinding time and speed can be obtained. It can be seen that for the same product particle size, the grinding efficiency decreases continuously with the increase in rotation speed at the same time.

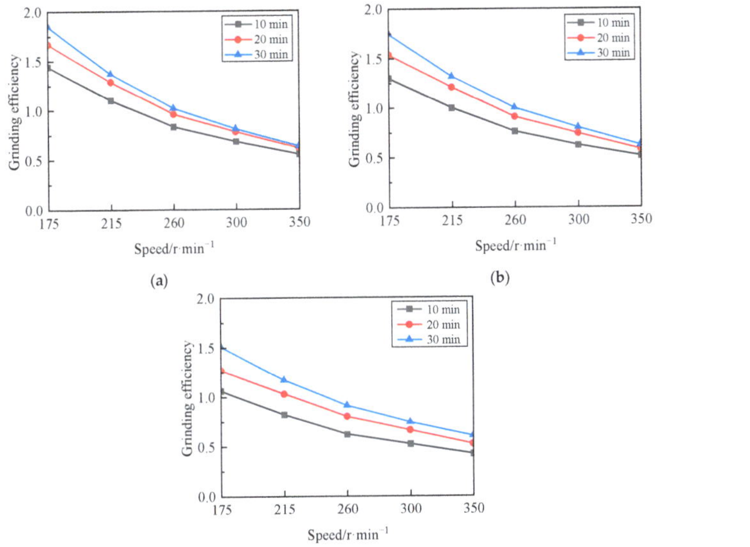

Figure 8. The variation law of grinding efficiency with rotational speed: (**a**) −45 μm; (**b**) −38 μm; (**c**) −28 μm.

Taking the −45 μm particle size after 10 min of grinding as an example, the changes in energy consumption, percentage of particle size, and grinding efficiency with respect to rotational speed were comprehensively compared. The energy consumption, percentage of particle size, and grinding efficiency under this setup were normalized to [0,1], as shown in Table 9. Their variation with rotational speed is shown in Figure 9.

Table 9. The normalized energy consumption, percentage of particle size, and grinding efficiency for −45 μm.

Number	A10	B10	C10	D10	E10
Energy consumption	0.31	0.44	0.61	0.77	1.00
Percentage of particle size	0.82	0.88	0.93	0.95	1.00
Grinding efficiency	1.00	0.76	0.58	0.47	0.38

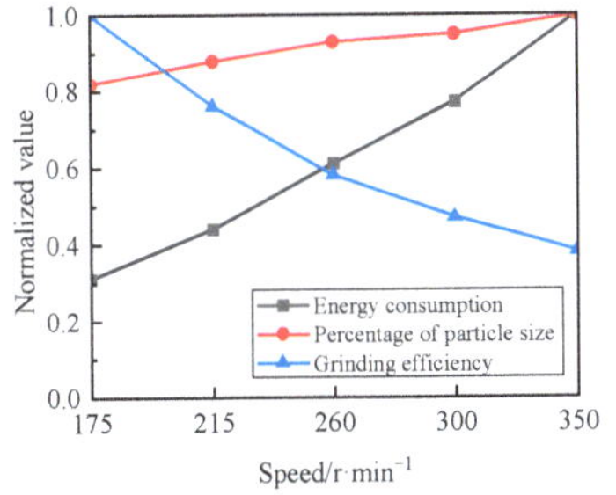

Figure 9. The variation law of energy consumption, percentage of particle size, and grinding efficiency with rotational speed.

It can be seen that with the increase in time, both energy consumption and percentage of particle size continue to increase with the increase in speed, and the growth rate of energy consumption is faster than that of the percentage of particle size. But the grinding efficiency decreases continuously with the increase in rotational speed. If high processing capacity is pursued within a certain period of time, high speed can be chosen, but it will result in energy loss. On the contrary, the low speed can be chosen if considering the grinding economy.

There is an intersection point between grinding efficiency and energy consumption in Figure 9, which is the critical speed selection point, and the corresponding agitator speed is around 260 rpm. If the speed is lower than this, the energy consumption of the mill is relatively small and the grinding efficiency is high. On the contrary, if the speed is higher than this, the energy consumption of the mill is relatively high, and the grinding efficiency is low.

5. Conclusions

The rotational speed of the agitator is one of the important parameters that affect the grinding efficiency of the vertical stirred mill. The effects of different speeds on energy consumption, product particle size, and grinding efficiency are analyzed in this study. An experimental vertical stirred mill with 2.2 kW operating power was used to grind iron ore. Five different speed parameters (175, 215, 260, 300, 350 rpm) are set. The torque of the agitator and product particle size are recorded every 10 min. Some important results and conclusions have been obtained.

Firstly, the energy consumption of the mill at different speeds and times is measured and calculated. At the same rotational speed, the energy consumption increases linearly with time, and the energy consumption also increases continuously as the rotational speed increases during the same time period.

Then, the particle percentages of −45, −38, −28 μm and the particle size at P_{80} in the samples are studied at different speeds and times. It was found that at the same rotational speed, the percentage of the same product particle size in the screened material continues to increase as the grinding time increases. It also increases as the agitator speed increases under the same grinding time. In addition, the particle size of P_{80} continuously decreases with the increase in grinding time at the same rotational speed, or the agitator speed increases under the same grinding time. After grinding at different speeds for a certain period of time, the energy consumption for the particle size to reach the same percentage is measured. It was found that the energy consumption increases with the increase in rotational speed when the particle size of the same product reaches the same proportion. From the above analysis, it can be concluded that increasing the rotational speed will increase the grinding effect, but it will trigger more energy consumption. The evaluation of grinding efficiency requires comprehensive consideration of grinding time, energy consumption, and product particle size.

Next, a new evaluation index to comprehensively reflect the grinding efficiency of the mill is proposed in this study. The grinding efficiency η can be defined as the ability of a mill to grind the same product per unit of time and energy consumption, which is a dimensionless evaluation index, and the larger its value, the higher the grinding efficiency. It was found that for the same product particle size, the grinding efficiency decreases continuously with the increase in rotation speed at the same time.

Finally, the changes in energy consumption, percentage of particle size, and grinding efficiency with respect to rotational speed are comprehensively compared. It was found that with the increase in time, both energy consumption and the percentage of particle size continue to increase with the increase in speed, and the growth rate of energy consumption is faster than that of the percentage of particle size, but the grinding efficiency decreases continuously with the increase in rotational speed. If high processing capacity is pursued within a certain period of time, high speed can be chosen, but it will result in energy loss. On the contrary, the low speed can be chosen if considering the grinding economy. In future research, different ore materials and feed particle sizes should be considered as evaluation indicators of grinding efficiency, which will improve the universality of the results.

Author Contributions: Conceptualization, Z.W. and B.T.; methodology, B.T. and X.S.; validation, B.C. and H.J.; investigation, B.T. and B.C.; data curation, H.J. and Y.L.; writing—original draft preparation, B.T. and H.J.; writing—review and editing, Z.W. and Y.L.; visualization, X.S.; supervision, B.T. and X.S.; project administration, B.C.; funding acquisition, B.T. and B.C. All authors have read and agreed to the published version of the manuscript.

Funding: This work is supported by the "Smart Mining Empowerment" of Major Scientific Research Project of CITIC Group (22C01).

Data Availability Statement: Data will be made available on request.

Acknowledgments: The authors wish to thank the peer reviewers for comments that significantly contributed to the paper.

Conflicts of Interest: Authors Biliang Tang, Xianzhou Song and Zhaohua Wang were employed by the CITIC Heavy Industries Co., Ltd., and Luoyang Mining Machinery Engineering Design Institute Co., Ltd. The remaining authors declare that the research was conducted in the absence of any commercial or financial relationships that could be construed as a potential conflict of interest.

References

1. Rhymer, D.; Ingram, A.; Sadler, K.; Windows-Yule, C.R.K. A discrete element method investigation within vertical stirred milling: Changing the grinding media restitution and sliding friction coefficients. *Powder Technol.* **2022**, *410*, 117825. [CrossRef]
2. Kumar, A.; Sahu, R.; Tripathy, S.K. Energy-Efficient Advanced Ultrafine Grinding of Particles Using Stirred Mills—A Review. *Energies* **2023**, *16*, 5277. [CrossRef]
3. Esteves, P.M.; Mazzinghy, D.B.; Galéry, R.; Machado, L.C.R. Industrial Vertical Stirred Mills Screw Liner Wear Profile Compared to Discrete Element Method Simulations. *Minerals* **2021**, *11*, 397. [CrossRef]
4. Can, M.; Altun, O. Performance Comparison of the Vertical and Horizontal Oriented Stirred Mill: Pilot Scale IsaMill vs. Full-Scale HIGMill. *Minerals* **2023**, *13*, 315. [CrossRef]
5. Zheng, J.; Harris, C.C.; Somasundaran, P. A study on grinding and energy input in stirred media mills. *Powder Technol.* **1996**, *86*, 171–178. [CrossRef]
6. Ford, E.; Naudé, N. Investigating the effect on power draw and grinding performance when adding a shell liner to a vertical fluidised stirred media mill. *Miner. Eng.* **2021**, *160*, 106698. [CrossRef]
7. de Carvalho, R.M.; Oliveira, A.L.R.; Petit, H.A.; Tavares, L.M. Comparing modeling approaches in simulating a continuous pilot-scale wet vertical stirred mill using PBM-DEM-CFD. *Adv. Powder Technol.* **2023**, *24*, 104135. [CrossRef]
8. Blecher, L.; Kwade, A.; Schwedes, J. Motion and stress intensity of grinding beads in a stirred media mill. Part 1: Energy density distribution and motion of single grinding beads. *Powder Technol.* **1996**, *86*, 59–68. [CrossRef]
9. Cleary, P.W.; Cummins, S.J.; Sinnott, M.D.; Delaney, G.W.; Morrison, R.D. Advanced comminution modelling: Part 2-Mills. *Appl. Math. Model.* **2020**, *88*, 307–348. [CrossRef]
10. Fragnière, G.; Naumann, A.; Schrader, M.; Kwade, A.; Schilde, C. Grinding Media Motion and Collisions in Different Zones of Stirred Media Mills. *Minerals* **2021**, *11*, 185. [CrossRef]
11. Rhymer, D.; Ingram, A.; Sadler, K.; Windows-Yule, C.R.K. Segregation in binary and polydisperse stirred media mills and its role on grinding effectiveness. *Powder Technol.* **2024**, *443*, 119921. [CrossRef]
12. Altun, O.; Sert, T.; Altun, D.; Toprak, A.; Kwade, A. Scale-up of Vertical Wet Stirred Media Mill (HIGmill) via Signature Plots, Stress Analyses and Energy Equations. *Miner. Eng.* **2024**, *205*, 108460. [CrossRef]
13. Guner, G.; Seetharaman, N.; Elashri, S.; Mehaj, M.; Bilgili, E. Analysis of Heat Generation During the Production of Drug Nanosuspensions in a Wet Stirred Media Mill. *Int. J. Pharm.* **2022**, *624*, 122020. [CrossRef]
14. Sinnott, M.; Cleary, P.W.; Morrison, R. Analysis of stirred mill performance using DEM simulation: Part 1 -Media motion, energy consumption and collisional environment. *Miner. Eng.* **2006**, *19*, 1537–1550. [CrossRef]
15. Sinnott, M.D.; Cleary, P.W.; Morrison, R.D. Is media shape important for grinding performance in stirred mills? *Miner. Eng.* **2011**, *24*, 138–151. [CrossRef]
16. Mankosa, M.J.; Adel, G.T.; Yoon, R.H. Effect of operating parameters in stirred ball mill grinding of coal. *Powder Technol.* **1989**, *59*, 255–260. [CrossRef]
17. Rahal, D.; Erasmus, D.; Major, K. Knelson-Deswik milling technology: Bridging the gap between low and high speed stirred mills. In Proceedings of the 43rd Annual Meeting of the Canadian Mineral Processor, Ottawa, ON, Canada, 18–20 January 2011; pp. 557–587.
18. De Oliveira, A.L.R.; de Carvalho, R.M.; Tavares, L.M. Predicting the effect of operating and design variables in grinding in a vertical stirred mill using a mechanistic mill model. *Powder Technol.* **2021**, *387*, 560–574. [CrossRef]

19. Fadhel, H.B.; Frances, C. Wet batch grinding of alumina hydrate in a stirred bead mill. *Powder Technol.* **2001**, *119*, 257–268. [CrossRef]
20. Mankosa, M.J.; Adel, G.T.; Yoon, R.H. Effect of media size in stirred ball mill grinding of coal. *Powder Technol.* **1986**, *49*, 75–82. [CrossRef]
21. Mingwei, G.; Forssberg, E. Prediction of product size distributions for a stirred ball mill. *Powder Technol.* **1995**, *84*, 101–106.
22. Yang, R.Y.; Jayasundara, C.T.; Yu, A.B.; Curry, D. DEM simulation of the flow of grinding media in IsaMill. *Miner. Eng.* **2006**, *19*, 984–994. [CrossRef]
23. Fukui, S.; Tsunazawa, Y.; Hisatomi, S.; Granata, G.; Tokoro, C.; Okuyama, K.; Iwamoto, M.; Sekine, Y. Effect of Agitator Shaft Direction on Grinding Performance in Media Stirred Mill: Investigation Using DEM Simulation. *Mater. Trans.* **2018**, *59*, 488–493. [CrossRef]
24. Liu, Z.; Wang, S.; Chen, Z.; Guan, W.; Guo, J.; Wu, S.; Huang, Q. Electromechanical coupling characteristics analysis of vertical stirred mill based on ECS-MBD-DEM. *Powder Technol.* **2024**, *433*, 119245. [CrossRef]
25. Cleary, P.W.; Sinnott, M.; Morrison, R. Prediction of slurry transport in SAG mills using SPH fluid flow in a dynamic DEM based porous media. *Miner. Eng.* **2006**, *19*, 1517–1527. [CrossRef]
26. Jayasundara, C.T.; Yang, R.Y.; Yu, A.B. Discrete Particle Simulation of Particle Flow in a Stirred Mill: Effect of Mill Properties and Geometry. *Ind. Eng. Chem. Res.* **2011**, *51*, 1050–1061. [CrossRef]
27. Jayasundara, C.T.; Yang, R.Y.; Guo, B.Y.; Yu, A.B.; Govender, I.; Mainza, A.; Westhuizen, A.V.D.; Rubenstein, J. CFD-DEM modelling of particle flow in IsaMills-Comparison between simulations and PEPT measurements. *Miner. Eng.* **2011**, *24*, 181–187. [CrossRef]
28. Winardi, S.; Widiyastuti, W.; Septiani, E.L.; Nurtono, T. Simulation of solid-liquid flows in a stirred bead mill based on computational fluid dynamics (CFD). *Mater. Res. Express* **2018**, *5*, 054002. [CrossRef]
29. Jonsén, P.; Pålsson, B.I.; Häggblad, H. A novel method for full-body modelling of grinding charges in tumbling mills. *Miner. Eng.* **2012**, *33*, 2–12. [CrossRef]
30. Ndimande, C.B.; Cleary, P.W.; Mainza, A.N.; Sinnott, M.D. Using two-way coupled DEM-SPH to model an industrial scale Stirred Media Detritor. *Miner. Eng.* **2019**, *137*, 259–276. [CrossRef]
31. Ndimande, C.B. A Study of the Charge Structure and Energy Utilisation in a Stirred Media Detritor Using DEM-SPH. Ph.D. Thesis, University of Cape Town, Cape Town, South Africa, 2022.
32. Jonsén, P.; Hammarberg, S.; Pålsson, B.I.; Lindkvist, G. Preliminary validation of a new way to model physical interactions between pulp, charge and mill structure in tumbling mills. *Miner. Eng.* **2019**, *130*, 76–84. [CrossRef]
33. Larsson, S.; Pålsson, B.I.; Parian, M.; Jonsén, P. A novel approach for modelling of physical interactions between slurry, grinding media and mill structure in wet stirred media mills. *Miner. Eng.* **2020**, *148*, 106180. [CrossRef]
34. Larsson, S.; Rodríguez Prieto, J.M.; Heiskari, H.; Jonsén, P. A Novel Particle-Based Approach for Modeling a Wet Vertical Stirred Media Mill. *Minerals* **2021**, *11*, 55. [CrossRef]
35. Böttcher, A.C.; Schilde, C.; Kwade, A. Experimental assessment of grinding bead velocity distributions and stressing conditions in stirred media mills. *Adv. Powder Technol.* **2021**, *32*, 413–423. [CrossRef]
36. Hacıfazlıoğlu, H.; Korkmaz, A.V. Performance comparison of stirred media mill and ball (BOND) mill in bauxite grinding. *Part. Sci. Technol.* **2020**, *38*, 404–408. [CrossRef]
37. Zhao, Z.; Song, R.; Yan, P.; Ren, S.; Wang, Y. Polygonal concave wear and irregular packing behavior of grinding balls for vertical grinding mill. *Miner. Eng.* **2023**, *200*, 108155. [CrossRef]
38. *GB/T 6730.71*; Iron Ores-Determination of Acid-Soluble Iron Content-Titrimetric Method. Standardization Administration of China: Beijing, China, 2014. (In Chinese)
39. *GB/T 6730.73*; Iron Ores-Determination of Total Iron Content-EDTA Photometric Titration Method. Standardization Administration of China: Beijing, China, 2016. (In Chinese)
40. Electrical Induction Motors-Torque vs. Speed. Available online: https://www.engineeringtoolbox.com/electrical-motors-torques-d_651.html (accessed on 15 November 2024).

minerals

MDPI

Article

Proposal of a Method for Calculating the Bond Work Index for Samples with Non-Standard Feed Particle Size Distribution

Vladimir Nikolić [1], Jesus Medina Pierres [2], Maria Sanchez Calvo [2], Juan M. Menéndez-Aguado [2,*], Milan Trumić [1], Maja S. Trumić [1] and Vladan Milošević [1]

[1] Technical Faculty in Bor, University of Belgrade, 19210 Bor, Serbia; vnikolic@tfbor.bg.ac.rs (V.N.); mtrumic@tfbor.bg.ac.rs (M.T.); majatrumic@tfbor.bg.ac.rs (M.S.T.); vmilosevic@tfbor.bg.ac.rs (V.M.)

[2] Asturias Raw Materials Institute, University of Oviedo, Gonzalo Gutiérrez Quirós, 33600 Mieres, Spain; medinajesus@uniovi.es (J.M.P.); sanchezcmaria@uniovi.es (M.S.C.)

* Correspondence: maguado@uniovi.es

Abstract: Determining the Bond grindability test in a ball mill is one of the most commonly used methods in the mining industry for measuring the hardness of ores. The test is an essential part of the Bond work index methodology for designing and calculating the efficiency of mineral grinding circuits. The Bond ball mill grindability test has several restrictions, including the sample's initial particle size distribution (PSD). This paper presents a method for calculating the Bond work index when the Bond ball mill grindability test is performed on samples with non-standard PSD. The presented equation includes a correction factor (k) and is applicable only for P_{100} = 75 μm. The defined method is then compared with methods proposed by other researchers, and conclusions are drawn as to which method results in less deviation. The presented model resulted in a mean square error of 0.66%.

Keywords: grindability; comminution; Bond work index; energy efficiency

1. Introduction

To determine the required grinding power, Fred Bond [1] developed an approach that equipment manufacturers still use as the state-of-the-art design methodology [2]. The method's theoretical basis is Bond's Third Theory of Comminution, which introduced the Bond work index (w_i). The numerical input of w_i describes the energy (kWh/t) required for reducing ore size from an initial very large size to a specific particle size (e.g., 80% of the product passes through a 100 μm sieve) [3,4].

Notwithstanding its widespread use, the standard Bond test has certain limitations [5,6]. The test requires a specific sample preparation and a precisely defined initial sample PSD. A sample weighing 10 kg and, usually, the help of a professional technician is needed to perform one Bond test, which can last more than 8 hours. Based on the standard Bond test, w_i is determined by a simulation of batch dry grinding in a closed cycle in a Bond ball mill until a circulating mill charge of 250% is obtained [1,7,8]. The w_i is calculated after the mill product's PSD.

These limitations prompted the proposal of alternative and faster methods and procedures for the Bond test simulation to determine the Bond w_i [9,10]. Alternative procedures presented by [11–16], faster procedures suggested by [17–21], and procedures for the Bond test simulation introduced by [22–26] were offered to shorten the Bond procedure duration. The procedures in [27–35] allowed estimating the w_i in other laboratory ball mills when the Bond laboratory mill was unavailable. All these procedures were based on Bond's grinding theory and comparison with the determination of the work index. The authors [12,36,37] suggested that decreased closing screen size increased the Bond w_i. Josefin and Doll (2018) [38] introduced an equation that could correct the w_i obtained for a sample calculated with a P_{80} value when the grinding operation is performed to a different P_{80} using a calibration sample. Even though these procedures exist, the knowledge of the Bond standard test is constantly extended by new research [9,25,34,39–45].

Magdalinovic et al. [44] determined the Bond w_i on samples of non-standard PSD. The Bond test was performed on dolomite, copper ore, and quartzite samples of different size classes (−3.327 + 0; −2.356 + 0; −1.651 + 0; −1.168 + 0; −0.833 + 0) mm. Based on the obtained results, the rules for changing the parameters G [g/rev], P_{80} [μm], and F_{80} [μm] in the Bond equation were established. For calculating the Bond work index using this method, the authors have proposed three equations. The maximum error obtained by the authors Magdalinovic et al. [44] using this method was less than 5%. Nikolić and Trumić [42] provided the procedure for determining the Bond w_i for finer samples. The paper presents a method for assessing Bond's w_i on samples showing a non-standard PSD.

The objective of this study was to demonstrate the theoretical and practical contribution to understanding the comminution process of non-standard particle size samples in a laboratory Bond ball mill. The research conducted in this study focused on the following:

- Monitoring the influence of the initial particle size of the sample on the Bond Work Index values during the execution of the standard Bond test;
- Defining a model for determining the Bond Work Index for standard particle size samples based on the known Bond Work Index value for a non-standard particle size sample;
- Testing the accuracy of the model using results from laboratory experiments and available data from the reviewed literature.

2. Materials and Methods

The samples used in this study were prepared with crushing in a jaw crusher and then sieved through a sieve with an opening size of −3.35 mm. Three mono-mineral-like ores were used in this study: zeolite, dacite, and basalt. The zeolite sample was taken from the "Jablanica" deposit near Kruševac [46]. The Jablanica zeolitized tuff deposit is located on the outskirts of the village of Jablanica, in the central part of the exploration area. The dacite sample was collected from the "Krš" open-pit mine near Ljubovija, which is a deposit of technical dacite stone of eruptive origin. The basalt sample was taken from the "Vrelo" basalt deposit, located in the vicinity of the village of Štava on the southeastern slopes of the central part of Mount Kopaonik. The samples had different resistance to comminution. Five 10 kg samples with different initial sizes (−3.35 + 0 mm; −2.36 + 0 mm; −1.70 + 0 mm; −1.18 + 0 mm; −0.850 + 0 mm) were formed for each type of raw material for a grinding test based on the standard Bond procedure. The PSDs of zeolite, dacite, and basalt are in Tables A1–A3 and Figures 1–3. Tables A1–A3 are provided in Appendix A.

A standard Bond laboratory mill, with a grinding chamber 305 × 305 mm and a rotation speed of 70 rev/min, was used to determine the Bond work index. The mill was filled with 15.5 to 30.6 mm in diameter balls, and the total weight was 20.125 kg. Dry grinding was used, simulating a closed grinding cycle until a circulating load of 250% was established [7]. A 75 μm closing screen size (P_{100}) was used. The Bond work index was calculated with Equation (1).

$$w_i = 1.1 \cdot \frac{44.5}{{P_{100}}^{0.23} \cdot G^{0.82} \cdot \left(\frac{10}{\sqrt{P_{80}}} \cdot \frac{10}{\sqrt{F_{80}}} \right)} [\frac{kWh}{t}] \quad (1)$$

wherein

P_{100}—closing screen size (μm);

G—net mass of undersize product per unit revolution of the mill, in g/rev;

P_{80}—the 80% passing product particle size (μm);

F_{80}—the 80% passing feed particle size (μm).

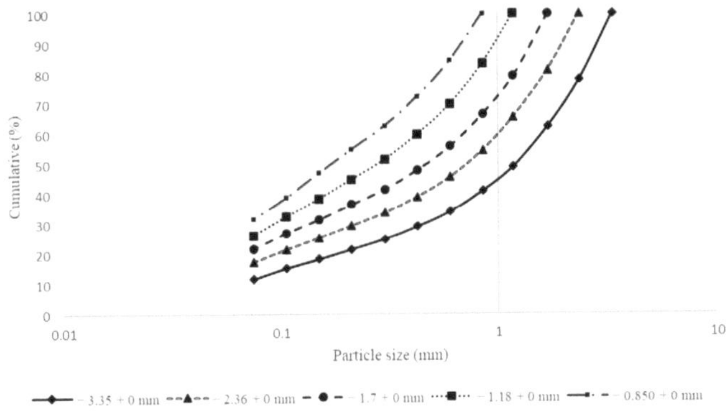

Figure 1. Zeolite sample PSDs.

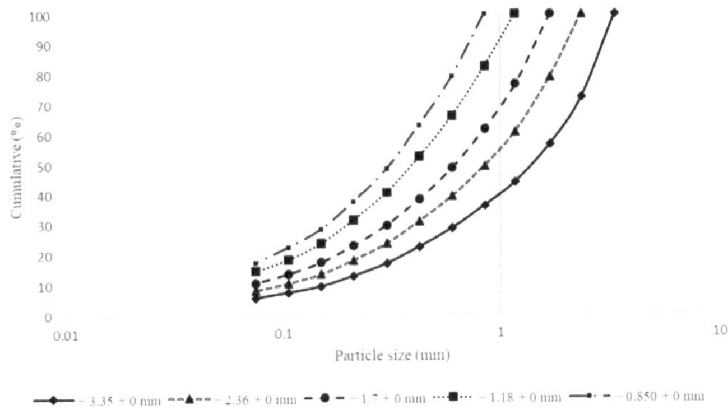

Figure 2. Dacite sample PSDs.

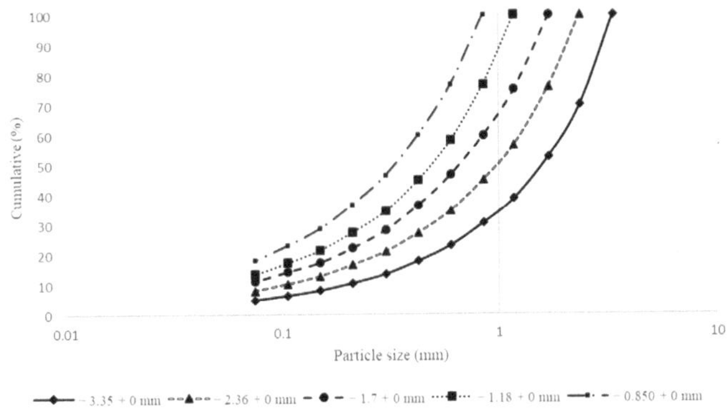

Figure 3. Basalt sample PSDs.

3. Results and Discussion

A single grindability test was performed on all samples, with no repetition of measurements. Therefore, the obtained results do not account for any potential experimental error that may arise from repeated measurements.

The obtained results for the Bond w_i for samples of zeolite, dacite, and basalt on non-standard size classes are shown in Table A4 and Figure 4. Table A4 is also included in Appendix A. The Bond work index was calculated using Equation (1) on samples with a standard particle size (F_{100} = 3.35 mm) and non-standard particle size ($F_{100} \leq 3.35$ mm). A closing screen size (P_{100}) of 75 μm was used for all samples.

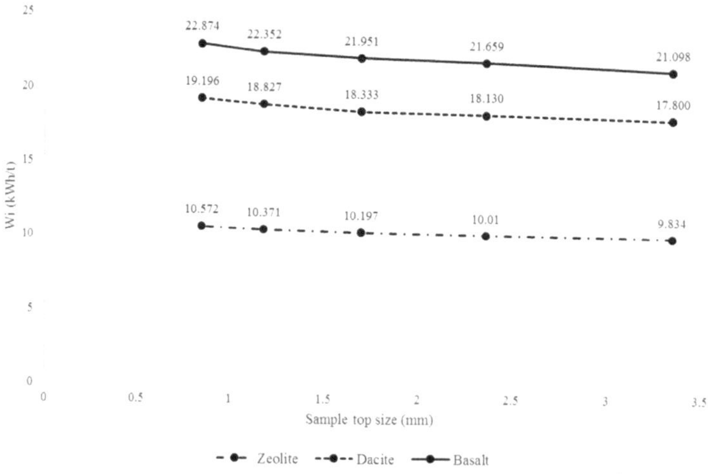

Figure 4. Values of Bond w_i with different feed PSD.

Based on the obtained results, it can be concluded that the Bond work index increases as the initial particle size of the sample decreases. The resistance of the raw material to comminution increases with decreasing particle size [44], which is one of the possible reasons for this phenomenon. An additional problem might be that feeds contain an excessive amount of finished product, which prolongs the attainment of a steady-state condition [30]. For practical purposes, a sample that contains 15% of the finished fraction or less is ideal [27]. However, in industrial practice, such a bulk material that has a higher ratio of fine fraction is not rare [30].

It is not unusual to receive at the laboratories samples to assess their w_i with a feed PSD differing from the standard PSD prescribed by Bond. The question arises as to whether it is possible to determine the Bond w_i if we obtain a sample of non-standard size. The presented method allows us to estimate the work index for a sample of standard size, although we do not have a sample of the size required by Bond. If we receive a sample of non-standard size, and we want to calculate the Bond w_i for a sample of standard PSD, the procedure is as follows:

1. First, Bond's standard test is performed on a sample of non-standard PSD;
2. Then, the obtained parameters ($w_{i,ns}$ and F_{ns}) from the Bond test on the non-standard PSD sample are used to estimate the w_i for the standard PSD sample, using Equation (2).

$$w_{i,c} = \frac{w_{i,ns} \cdot {F_{ns}}^{0.05}}{k} \quad (2)$$

where

$w_{i,ns}$—work index for a sample of non-standard PSD [kWh/t];
$w_{i,c}$—calculated work index for a sample of standard PSD [kWh/t];
F_{ns}—the 80% passing sample of non-standard PSD [μm];
k—coefficient that depends on the ore grindability (Table 1).

Table 1. Value of coefficient k [42].

w_i [kWh/t]	10–17	18–20	>21
k	1.47	1.48	1.49

Equation (2) was tested on samples of varying grindability, including zeolite, dacite, and basalt. The aim was to test the validity of the equation and check its accuracy and reliability. The results of testing Equation (2) on the samples of zeolite, dacite, and basalt are presented in Table 2. A search of the literature presented papers where researchers determined the Bond work index using samples of non-standard sizes. These results are added to the examination to confirm the validity and accuracy of the presented methodology for data not included in the empirical "training data" used to derive the model. In the continuation of the paper, the equation was tested on all other available data found in the literature.

Table 2. Results obtained using Equation (2).

Sample	Class Size (mm)	F_{ns} (μm)	P_{100} = 75 μm $w_{i,ns}$ (kWh/t)	k	w_i (kWh/t)	$w_{i,c}$ (kWh/t)	Δ (%)
Zeolite	−2.36 + 0	1652	10.010	1.47	9.834	9.863	−0.29
	−1.70 + 0	1090	10.197			9.841	−0.07
	−1.18 + 0	727	10.371			9.827	+0.07
	−0.850 + 0	544	10.572			9.854	−0.20
Dacite	−2.36 + 0	1729	18.130	1.48	17.800	17.784	+0.09
	−1.70 + 0	1253	18.333			17.696	+0.58
	−1.18 + 0	807	18.827			17.777	+0.13
	−0.850 + 0	609	19.196			17.873	−0.41
Basalt	−2.36 + 0	1800.1	21.659	1.49	21.098	21.145	−0.22
	−1.70 + 0	1278	21.951			21.067	+0.15
	−1.18 + 0	892	22.352			21.069	+0.14
	−0.850 + 0	633	22.874			21.195	−0.46

The presented results show the Bond work index value that would be obtained for a sample with standard particle size (F_{100} = 3.35 mm) if a sample with non-standard particle size ($F_{100} \leq$ 3.35 mm) were present in the initial sample. Based on the results obtained using Equation (2), an error of less than 1% is obtained. Few researchers have addressed this issue because the preparation of samples and testing of the Bond test takes a very long time and usually requires the help of expert technical personnel. Equation (2) was tested on available data that could be found in the literature, and the results are shown in Table 3.

The results in Table 3 show that the mean square error is 0.66% when Equation (2) is used to calculate the w_i for a standard PSD sample. The application of Equation (2) in Table 3 was only tested for a 75 μm closing screen size. The presented method for determining the w_i for a sample with standard PSD when the value of the Bond w_i for a sample with non-standard particle size is known can only be applied when the Bond w_i is determined at P_{100} = 75 μm. In their study, the authors Magdalinovic et al. [44] presented a method for determining the Bond ($w_{i,cM}$) for a sample with a standard PSD, where the value of the Bond w_i is known for a sample with a non-standard PSD and which can be used with a P_{100} = 75 μm and P_{100} = 149 μm. The idea was to compare these two methods, but only at a closing screen size of 75 μm, because the presented method is applicable only at a closing screen size of 75 μm, and to determine which model leads to a lower deviation. The comparative results obtained with Equation (2) and the model presented by Magdalinovic et al. in [44] are shown in Table 4, respectively. Table 5 presents the results obtained when using the model by Magdalinovic et al. in [44] for a closing screen size of 149 μm.

Table 3. Comparative results of w_i obtained by Bond test and Equation (2).

Sample	Ref.	Class Size (mm)	F_{ns} (μm)	P_{100} = 75 μm		w_i (kWh/t)	$w_{i,c}$ (kWh/t)	Δ (%)	$Δ^2$
				$w_{i,ns}$ (kWh/t)	k				
Zeolite		−2.36 + 0	1652	10.01			9.86	−0.29	0.084
		−1.70 + 0	1090	10.20			9.84	−0.07	0.005
		−1.18 + 0	727	10.37		9.834	9.83	+0.07	0.0049
		−0.850 + 0	544	10.57			9.85	−0.20	0.040
Dolomite	[44]	−2.356 + 0	1662	12.91	1.47		12.72	−0.16	0.026
		−1.651 + 0	1090	13.16			12.70	0.00	0.000
		−1.168 + 0	727	13.38		12.70	12.65	+0.39	0.152
		−0.833 + 0	544	13.69			12.76	−0.47	0.221
Cu ore	[44]	−2.356 + 0	1729	15.70			15.51	+1.02	1.040
		−1.651 + 0	1253	15.84			15.39	+1.79	3.204
		−1.168 + 0	807	16.19		15.67	15.39	+1.79	3.204
		−0.833 + 0	609	16.79			15.74	−0.45	0.202
Dacite		−2.36 + 0	1729	18.13			17.78	+0.09	0.008
		−1.70 + 0	1253	18.33			17.70	+0.58	0.336
		−1.18 + 0	807	18.83	1.48	17.80	17.78	+0.13	0.017
		−0.850 + 0	609	19.20			17.87	−0.41	0.168
Basalt		−2.36 + 0	1800.1	21.66			21.14	−0.22	0.048
		−1.70 + 0	1278	21.95			21.07	+0.15	0.022
		−1.18 + 0	892	22.35		21.10	21.07	+0.14	0.020
		−0.850 + 0	633	22.87			21.20	−0.46	0.212
Quartzite	[44]	−2.356 + 0	1790	23.17	1.49		22.61	+0.09	0.001
		−1.651 + 0	1240	23.52			22.54	+0.40	0.160
		−1.168 + 0	870	24.14		22.63	22.73	−0.44	0.194
		−0.833 + 0	610	24.72			22.86	−1.02	1.040
				Sum					10.4173
				Mean square error $\sqrt{\frac{\Delta^2}{N=24}}$					0.66

The results in Tables 4 and 5 show that when testing the model presented by Magdalinovic et al. [44], the mean square error is 2.78% for a 75 μm closing screen size and 1.80% for a 149 μm closing screen size. Based on the results obtained, it can be concluded that when determining the w_i for a standard PSD sample, knowing the w_i for a non-standard PSD sample on a 75 μm closing screen size, Equation (2) poses a lower deviation than the model presented in [44]. Therefore, when determining the Bond w_i for a standard PSD sample, and knowing the Bond w_i for a non-standard particle size sample on a P_{100} = 75 μm closing screen size, it should be recommended the use of Equation (2), while for a coarser grinding, P_{100} = 149 μm, the model of Magdalinovic et al. [44] can be used, for it is currently the only model tested at this closing screen size.

Table 4. Comparative results of the w_i obtained by the Bond test and the method in [44] (P_{100} = 75 μm).

Sample	Ref.	Class Size (mm)	F_{ns} (μm)	$w_{i,ns}$ (kWh/t)	w_i (kWh/t)	$w_{i,cM}$ (kWh/t)	Δ (%)	$Δ^2$
Zeolite		−2.36 + 0	1652	10.01	9.83	9.93	−0.98	0.96
		−1.70 + 0	1090	10.20		10.14	−3.11	9.67
		−1.18 + 0	727	10.37		10.12	−2.91	8.47
		−0.850 + 0	544	10.57		9.79	+0.45	0.20
Dolomite	[44]	−2.356 + 0	1662	12.91	12.70	12.91	−1.65	2.72
		−1.651 + 0	1090	13.16		13.04	−2.68	7.18
		−1.168 + 0	727	13.38		12.94	−1.89	3.57
		−0.833 + 0	544	13.69		12.85	−1.18	1.39
Cu ore	[44]	−2.356 + 0	1729	15.70	15.67	15.69	−0.13	0.02
		−1.651 + 0	1253	15.84		15.69	−0.13	0.02
		−1.168 + 0	807	16.19		15.72	−0.32	0.10
		−0.833 + 0	609	16.79		15.78	−0.70	0.49
Dacite		−2.36 + 0	1729	18.13	17.80	18.09	−1.63	2.66
		−1.70 + 0	1253	18.33		18.19	−2.19	4.80
		−1.18 + 0	807	18.83		18.35	−3.09	9.55
		−0.850 + 0	609	19.20		18.81	−5.67	32.15
Basalt		−2.36 + 0	1800.1	21.66	21.10	21.52	−2.00	4.00
		−1.70 + 0	1278	21.95		21.85	−3.56	12.67
		−1.18 + 0	892	22.35		22.05	−4.51	20.34
		−0.850 + 0	633	22.87		21.90	−3.80	14.44
Quartzite	[44]	−2.356 + 0	1790	23.17	22.63	23.16	−2.34	5.48
		−1.651 + 0	1240	23.52		23.39	−3.36	11.29
		−1.168 + 0	870	24.14		23.64	−4.46	19.89
		−0.833 + 0	610	24.72		23.46	−3.67	13.47
			Sum					185.53
			Mean square error $\sqrt{\frac{\Delta^2}{N=24}}$					2.78

Table 5. Comparative results of the w_i obtained by the Bond test and the method in [44] (P_{100} = 149 μm).

Sample	Ref.	Class Size (mm)	F_{ns} (μm)	$w_{i,ns}$ (kWh/t)	w_i (kWh/t)	$w_{i,cM}$ (kWh/t)	Δ (%)	$Δ^2$
Dolomite	[44]	−2.356 + 0	1662	9.77	9.82	9.56	+2.65	7.023
		−1.651 + 0	1090	10.59		9.91	−0.92	0.846
		−1.168 + 0	727	11.44		9.94	−1.22	1.488
		−0.833 + 0	544	12.19		9.83	−0.10	0.010
Cu ore	[44]	−2.356 + 0	1729	15.86	15.32	15.43	−0.72	0.518
		−1.651 + 0	1253	16.46		15.43	−0.72	0.518
		−1.168 + 0	807	17.42		15.33	−0.07	0.005
		−0.833 + 0	609	18.93		15.11	+1.37	1.877
Quartzite	[44]	−2.356 + 0	1790	18.92	19.00	18.61	−2.05	4.203
		−1.651 + 0	1240	19.13		18.25	+3.95	15.603
		−1.168 + 0	870	20.61		18.67	+1.74	3.028
		−0.833 + 0	610	23.30		19.37	−1.95	3.803
			Sum					38.922
			Mean square error $\sqrt{\frac{\Delta^2}{N=25}}$					1.80

4. Conclusions

The determination of the Bond work index is considered the most popular method used for calculating critical parameters of the grinding process, selecting equipment, and controlling the grinding process. The Bond test is regarded as a standard laboratory procedure for determining the parameter w_i. As an initial condition for conducting the Bond grindability test, the upper particle size limit of the raw material must be 3.35 mm (F_{100} = 3.35 mm).

The determination of the Bond work index in a ball mill for raw materials with an initial particle size smaller than the standard size (−3.35 + 0 mm) has not received sufficient attention. Therefore, this study investigates the effect of raw material particle size, with an upper size limit smaller than 3.35 mm, on the Bond work index value. The experiments were conducted as a function of the initial sample size on raw materials with different mineral compositions and physical-mechanical properties (zeolite, dacite, and basalt).

Based on the experimentally obtained results from tests conducted on samples of zeolite, dacite, and basalt, as well as the analysis and discussion of the results, the following conclusions were drawn:

- The value of the Bond work index increases as the particle size of the raw material decreases. For samples with an upper size limit significantly smaller than the standard size, the Bond work index value is higher than the Bond work index value obtained for the standard-sized sample.
- A model for determining the Bond work index for standard-sized samples has been presented, provided the Bond work index for a non-standard-sized sample is known.
- The model is applicable only for a closing screen size (P_{100}) of 75 μm.
- The presented model was then tested on all available data found in the literature to verify the accuracy and validity of the model.

From the information presented, it can be concluded that the Bond work index for a standard particle size sample can be determined if the value of the Bond work index for a non-standard particle size sample is known. Applying Equation (2) to the tested samples to obtain wi resulted in a mean square error of −0.66%, which is within the limits of reproducibility of the standard Bond test. The results obtained with Equation (2) were compared with the results of Magdalinovic et al. (2012) [44], leading to the following conclusions:

- When determining the Bond work index for a standard particle size sample, if the Bond work index value for a non-standard particle size sample on a 75 μm closing screen size is known, Equation (2) should be used, as it provides more reliable results.
- When determining the Bond work index for a standard particle size sample, if the Bond work index value for a non-standard particle size sample on a closing screen size of 149 μm is known, the model of Magdalinovic et al. [44] should be used, as it is currently the only model tested on a closing screen size of 149 μm.

The low mean square error confirms the accuracy and validity of the presented method for determining the Bond work index on a standard particle size sample when the Bond work index value is known for a non-standard particle size sample. The presented method represents a practical contribution that will significantly help practitioners when planning a new plant or optimizing an existing one.

Author Contributions: Conceptualization, V.N., J.M.M.-A. and M.T.; methodology, V.N., M.S.C. and J.M.P.; investigation, V.N., M.S.C., J.M.P. and V.M.; resources, V.N., M.S.C. and J.M.P.; formal analysis, M.S.C., J.M.P. and V.M.; writing—original draft preparation, V.N., M.S.C. and J.M.P.; writing—review and editing, validation, M.S.T., V.N., J.M.M.-A. and V.M.; supervision, J.M.M.-A.; funding acquisition, V.N. All authors have read and agreed to the published version of the manuscript.

Funding: The research presented in this paper had the financial support of the Ministry of Science, Technological Development and Innovation of the Republic of Serbia, within the funding of the scientific research work at the University of Belgrade, Technical Faculty in Bor, according to the contract with registration number 451-03-137/2025-03/200131. The Asturias Raw Materials Institute members' contribution was partially funded by the Horizon Europe project 101137932 Li4life.

Data Availability Statement: The original contributions presented in this study are included in the article. Further inquiries can be directed to the corresponding author.

Conflicts of Interest: The authors declare no conflicts of interest.

Appendix A

Table A1. Zeolite sample PSDs.

Particle Size (mm)	Class Size in mm (%)				
	−3.35 + 0	−2.36 + 0	−1.70 + 0	−1.18 + 0	−0.850 + 0
−3.35 + 2.36	21.95				
−2.36 + 1.70	15.55	18.74			
−1.70 + 1.18	13.35	16.72	20.64		
−1.18 + 0.850	8.10	9.99	12.74	16.42	
−0.850 + 0.600	6.90	8.86	10.69	13.64	15.51
−0.600 + 0.425	4.95	6.51	7.80	9.99	12.02
−0.425 + 0.300	4.14	5.33	6.46	8.28	9.68
−0.300 + 0.212	3.37	4.26	5.02	6.66	7.82
−0.212 + 0.150	3.26	4.04	5.00	6.39	7.75
−0.150 + 0.106	3.01	3.87	4.56	6.07	8.23
−0.106 + 0.075	3.62	4.23	5.27	6.25	7.20
−0.075 + 0.00	11.80	17.45	21.82	26.30	31.79
Σ	100.00	100.00	100.00	100.00	100.00

Table A2. Dacite sample PSDs.

Particle Size (mm)	Class Size in mm (%)				
	−3.35 + 0	−2.36 + 0	−1.70 + 0	−1.18 + 0	−0.850 + 0
−3.35 + 2.36	27.77				
−2.36 + 1.70	15.60	21.08			
−1.70 + 1.18	12.59	18.24	23.42		
−1.18 + 0.850	7.99	11.28	14.83	17.54	
−0.850 + 0.600	7.52	10.18	12.98	16.40	20.91
−0.600 + 0.425	6.10	8.28	10.36	13.47	16.02
−0.425 + 0.300	5.45	7.39	8.76	11.94	14.62
−0.300 + 0.212	4.05	5.46	6.73	9.21	10.95
−0.212 + 0.150	3.39	4.57	5.55	7.76	9.25
−0.150 + 0.106	2.08	3.00	3.78	5.46	5.93
−0.106 + 0.075	1.77	2.39	2.92	3.68	5.02
−0.075 + 0.00	5.69	8.13	10.67	14.54	17.30
Σ	100.00	100.00	100.00	100.00	100.00

Table A3. Basalt sample PSDs.

Particle Size (mm)	Class Size in mm (%)				
	−3.35 + 0	−2.36 + 0	−1.70 + 0	−1.18 + 0	−0.850 + 0
−3.35 + 2.36	29.92				
−2.36 + 1.70	17.28	23.84			
−1.70 + 1.18	14.10	19.64	24.71		
−1.18 + 0.850	8.24	11.63	15.34	23.05	
−0.850 + 0.600	7.38	10.22	13.14	18.60	23.14
−0.600 + 0.425	5.18	7.39	10.20	13.29	16.76
−0.425 + 0.300	4.33	6.19	8.32	10.45	13.55
−0.300 + 0.212	2.96	4.36	6.00	7.00	9.71
−0.212 + 0.150	2.44	3.76	4.84	6.00	8.04
−0.150 + 0.106	1.72	2.59	3.10	4.23	5.64
−0.106 + 0.075	1.47	2.28	3.14	3.79	4.98
−0.075 + 0.00	4.98	8.10	11.21	13.59	18.18
Σ	100.00	100.00	100.00	100.00	100.00

Table A4. Parameters F_{80} and value of w_i for samples of zeolite, dacite and basalt used on non-standard size classes.

Sample	Class Size (mm)	P_{100} = 75 μm	
		F_{80} (μm)	w_i (kWh/t)
Zeolite	−3.35 + 0	2440	9.834
	−2.36 + 0	1652	10.010
	−1.70 + 0	1090	10.197
	−1.18 + 0	727	10.371
	−0.850 + 0	544	10.572
Dacite	−3.35 + 0	2646	17.800
	−2.36 + 0	1729	18.130
	−1.70 + 0	1253	18.333
	−1.18 + 0	807	18.827
	−0.850 + 0	609	19.196
Basalt	−3.35 + 0	2609.1	21.098
	−2.36 + 0	1800.1	21.659
	−1.70 + 0	1278	21.951
	−1.18 + 0	892	22.352
	−0.850 + 0	633	22.874

References

1. Bond, F.C. The third theory of comminution. *Trans. Am. Inst. Min. Metall. Eng.* **1952**, *193*, 484–494.
2. Napier-Munn, T.; Morell, S.; Morrison, R.; Kojovic, T. *Mineral Comminution Circuits: Their Operation and Optimisation*, 2nd ed.; Julius Kruttschnitt Mineral Research Centre, The University of Queensland: Queensland, Australia, 1999; p. 413.
3. Haffez, G.S.A. Correlation between bond work index and mechanical properties of some saudi ores. *J. Eng. Sci.* **2012**, *40*, 271–280.
4. Gupta, V.K. Understanding production of fines in batch ball milling for mill scale-up design using the population balance model. *Adv. Powder Technol.* **2018**, *29*, 2035–2047.
5. Mosher, J.B.; Tague, C.B. Conduct and precision of Bond grindability testing. *Miner. Eng.* **2001**, *14*, 1187–1197.
6. Jankovic, A.; Suthers, S.; Wills, T.; Valery, W. Evaluation of dry grinding using HPGR in closed circuit with an airclassifier. *Miner. Eng.* **2015**, *71*, 133–138.
7. Bond, F.C. Standard grindability test tabulated. *Trans. Am. Inst. Min. Metall. Eng.* **1949**, *183*, 313–329.
8. Bond, F.C. Crushing grinding calculation part I and II. *Br. Chem. Eng.* **1961**, *6*, 378–385, 543–548.
9. Nikolić, V.; García, G.G.; Coello-Velázquez, L.A.; Menéndez-Aguado, M.J.; Trumić, M.; Trumić, S.M. A Review of Alternative Procedures to the Bond Ball Mill Standard Grindability Test. *Metals* **2021**, *11*, 1114. [CrossRef]
10. Nikolić, V.; Ferradal, P.S.; Pierres, J.M.; Menéndez-Aguado, J.M.; Trumić, M. Methods for Estimating the Bond Work Index for Ball Mills. *Minerals* **2024**, *14*, 1264. [CrossRef]
11. Berry, T.F.; Bruce, R.W. A simple method of determining the grindability of ores. *Can. Min. J.* **1966**, *87*, 63–65.
12. Smith, R.W.; Lee, K.H. A comparison of data from Bond type simulated closed circuit and batch type grindability tests. *Trans. Metallur. Soc. AIME* **1968**, *241*, 91–99.
13. Kapur, P.C. Analysis of the Bond grindability test. *Trans. Inst. Min. Metallur.* **1970**, *79*, 103–107.
14. Horst, W.E.; Bassarear, J.H. Use of simplified ore grindability technique to evaluate plant performance. *Trans. Metallur. Soc. AIME* **1977**, *260*, 348–351.
15. Karra, V.K. Simulation of Bond grindability tests. *CIM Bull.* **1981**, *74*, 195–199.
16. Mular, A.L.; Jergensen, G.V. *Design and Installation of Comminution Circuits*; Society of Mining Engineers of the American Institute of Mining, Metallurgical, and Petroleum Engineers, Inc.: New York, NY, USA, 1982; p. 1022.
17. Magdalinović, N. A procedure for rapid determination of the Bond work index. *Int. J. Miner. Process.* **1989**, *27*, 125–132.
18. Magdalinović, N. Abbreviated test for quick determination of Bond's Work index. *J. Min. Metallur.* **2003**, *39*, 1–10.
19. Ahmadi, R.; Shahsavari, S. Procedure for determination of ball Bond work index in the commercial operations. *Miner. Eng.* **2009**, *22*, 104–106. [CrossRef]
20. Ford, E.; Sithole, V. A Comparison of Test Procedures for Estimating the Bond Ball Work Index on Zambian/DRC Copper-Cobalt Ores and Evaluation of Suitability for Use in Geometallurgical Studies. In Proceedings of the Copper Cobalt Africa, Incorporating the 8th Southern African Base Metals Conference, Livingstone, Zambia, 6–8 July 2015; pp. 65–68.
21. Todorovic, D.; Trumic, M.; Andric, L.; Milosevic, V.; Trumic, M. A quick method for bond work index approximate value determination. *Physicochem. Probl. Miner. Process.* **2017**, *53*, 321–332.
22. Lewis, K.A.; Pearl, M.; Tucker, P. Computer Simulation of the Bond Grindability test. *Miner. Eng.* **1990**, *3*, 199–206. [CrossRef]
23. Aksani, B.; Sönmez, B. Simulation of bond grindability test by using cumulative based kinetic model. *Miner. Eng.* **2000**, *13*, 673–677. [CrossRef]

24. Ciribeni, V.; Bertero, R.; Tello, A.; Puerta, M.; Avellá, E.; Paez, M.; Menéndez-Aguado, J.M. Application of the Cumulative Kinetic Model in the Comminution of Critical Metal Ores. *Metals* **2020**, *10*, 925.
25. Camalan, M. A computational algorithm coupled with a particle selection routine for the simulation of the Bond locked-cycle test. *Miner. Eng.* **2022**, *176*, 107345.
26. Rodríguez-Torres, I.; Tuzcu, E.T.; Andrade-Martínez, J.; Rosales-Marín, G. Estimation methodology for Bond ball mill work index experiment output via mathematical modeling. *Miner. Eng.* **2023**, *201*, 108186.
27. Yap, R.F.; Sepulveda, J.L.; Jauregui, R. Determination of the Bond work index using an ordinary batch ball mill. In *Design and Installation of Comminution Circuits*; Mular, A.L., Jergensen, G.V., Eds.; AIME: New York, NY, USA, 1982; pp. 176–203.
28. Nematollahi, H. New size laboratory ball mill for Bond work index determination. *Miner. Eng.* **1994**, *46*, 352–353.
29. Menéndez-Aguado, J.M.; Dzioba, B.R.; Coello-Valazquez, A.L. Determination of work index in a common laboratory mill. *Miner. Metall. Process.* **2005**, *22*, 173–176.
30. Mucsi, G. Fast test method for the determination of the grindability of fine materials. *Chem. Eng. Res. Des.* **2008**, *86*, 395–400. [CrossRef]
31. Swain, R.; Rao, R.B. Alternative Approaches for Determination of Bond Work Index on Soft Friable Partially Laterised Khondalite Rocks of Bauxite Mine Waste Materials. *J. Miner. Mater. Charact. Eng.* **2009**, *8*, 729–743.
32. Saeidi, N.; Noaparast, M.; Azizi, D.; Aslani, S.; Ramadi, A. A developed approach based on grinding time to determine ore comminution properties. *J. Miner. Environ.* **2013**, *4*, 105–112.
33. Mwanga, A.; Rosenkranz, J.; Lamberg, P. Development and experimental validation of the Geometallurgical Comminution Test (GCT). *Miner. Eng.* **2017**, *108*, 109–114.
34. Lvov, V.V.; Chitalov, L.S.; Struk, G.V.; Rakov, A.V. Research of the MSL-14K mill applicability to determine the Bond ball mill work index, MIAB. *Min. Informational Anal. Bull.* **2022**, *6*, 290–303. (In Russian) [CrossRef]
35. Arellano-Piña, R.; Sanchez-Ramirez, E.A.; Pérez-Garibay, R.; Gutiérrez-Pérez, V.H. Bond's work index estimation using non-standard ball mills. *Physicochem. Probl. Miner. Process.* **2023**, *59*, 172458.
36. Tüzün, M.A. Wet bond mill test. *Miner. Eng.* **2001**, *14*, 369–373.
37. Deniz, V.; Sütçü, N.; Umucu, Y. The effect of circulating load and test sieve size on the Bond work index based on natural amorphous silica. In Proceedings of the 18th International Mining Congress and Exhibition of Turkey-IMCET, Antalya, Turkey, 10–13 June 2003; pp. 517–522, ISBN 975-395-605-3.
38. Josefin, Y.; Doll, A.G. Correction of Bond Ball Mill Work Index Test for Closing Mesh Sizes, Procemin-Geomet 2018. In Proceedings of the 14th International Mineral Processing Conference & 5th International Seminar on Geometallurgy, Santiago, Chile, 28–30 November 2018; pp. 1–12.
39. Ciribeni, V.; Menéndez-Aguado, J.M.; Bertero, R.; Tello, A.; Avellá, E.; Paez, M.; Coello-Velázquez, A.L. Unveiling the Link between the Third Law of Comminution and the Grinding Kinetics Behaviour of Several Ores. *Metals* **2021**, *11*, 1079. [CrossRef]
40. García, G.G.; Oliva, J.; Guasch, E.; Anticoi, H.; Coello-Velázquez, A.L.; Menéndez-Aguado, J.M. Variability Study of Bond Work Index and Grindability Index on Various Critical Metal Ores. *Metals* **2021**, *11*, 970. [CrossRef]
41. García, G.G.; Coello-Velázquez, A.L.; Pérez, B.F.; Menéndez-Aguado, J.M. Variability of the Ball Mill Bond's Standard Test in a Ta Ore Due to the Lack of Standardization. *Metals* **2021**, *11*, 1606. [CrossRef]
42. Nikolić, V.; Trumić, M. A new approach to the calculation of bond work index for finer samples. *Miner. Eng.* **2021**, *165*, 106858.
43. Nikolić, V.; Doll, A.; Trumić, M. A new methodology to obtain a corrected Bond ball mill work index valid with non-standard feed size. *Miner. Eng.* **2022**, *188*, 107822.
44. Magdalinovic, N.; Trumic, M.; Trumic, G.; Magdalinovic, S.; Trumic, M. Determination of the Bond work index on samples of non-standard size. *Int. J. Miner. Process.* **2012**, *114-117*, 48–50.
45. Stamboliadis, E.T. A contribution to the relationship of energy and particle size in the comminution of brittle particulate materials. *Miner. Eng.* **2002**, *15*, 707–713.
46. Andrić, L.; Trumić, M.; Trumić, M.; Nikolić, V. Micronization of zeolite in vibration mill. *Recycl. Sustain. Dev.* **2018**, *11*, 63–71.

minerals

MDPI

Article

Accuracy and Precision of the Geopyörä Breakage Test

Marcos Bueno [1,*], Thiago Almeida [1], Leonardo Lara [1], Malcolm Powell [1] and Homero Delboni [2]

1 Geopyörä Oy, Toppilansaarentie 3 B39, 90510 Oulu, Finland; thiago@geopyora.com (T.A.); leonardo@geopyora.com (L.L.); cto@geopyora.com (M.P.)
2 Department of Mining and Petroleum Engineering, Polytechnical School, University of São Paulo, Av. Prof. Mello Moraes, 2373-Cidade Universitária, Sao Paulo 05508-900, SP, Brazil; hdelboni@usp.br
* Correspondence: ceo@geopyora.com

Abstract: The Geopyörä breakage test uses two counter-rotating wheels to nip and crush rock specimens with a tightly controlled gap between rollers. This paper presents the detailed measures conducted to evaluate the accuracy and precision of energy measurements across various ore types using the Geopyörä. Force measurement was assessed just for its precision. The outputs were compared directly to the drop weight test (DWT) measures of fragmentation at the same energy and fitted A and b parameters. Test reproducibility was evaluated using a Round-Robin methodology, testing several samples in multiple laboratories. The results confirmed that the new test has sufficient accuracy to match DWT results and excellent precision to assure reproducibility.

Keywords: comminution; ore breakage characterization; variability

1. Introduction

The Geopyörä breakage test device, which is pictured in Figure 1, was developed to fulfil a need in the industry for improved precision in the characterisation of an orebody. The method developed addresses this by enabling rapid measurements while measuring the input energy and force needed to fracture every particle [1]. The outcome is a single test that delivers multiple breakage parameters, bypassing the need for a suite of expensive tests. The target application is in orebody modelling for process performance prediction.

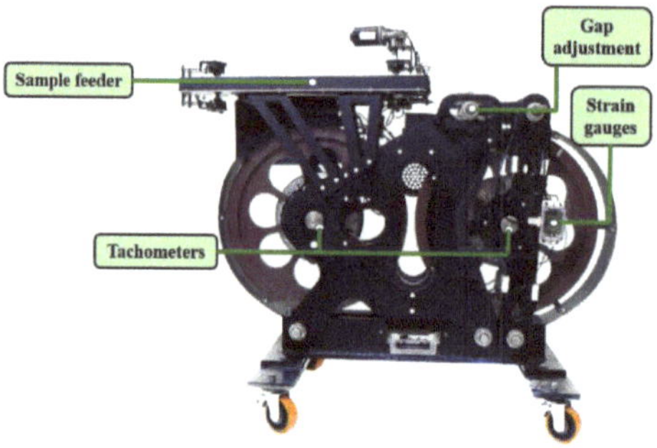

Figure 1. Geopyörä test equipment (original).

The concept behind the Geopyörä breakage test is to use counter-rotating wheels to nip and crush a rock with a tightly controlled reduction ratio from the feed to a defined gap between rollers, as illustrated in Figure 2. However, before delving into its mechanics and performance, the context of measurement precision compared to current standard test methods should be addressed.

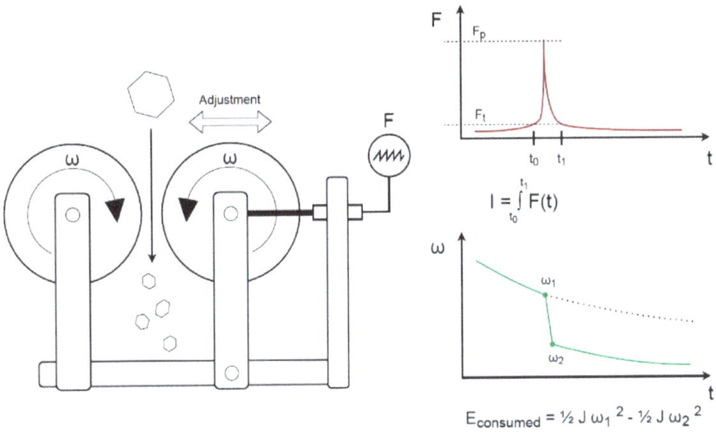

Figure 2. Double Wheel (Geopyörä) breakage test schematic of operation.

The drop weight test is an industry accepted standard test method used to characterise the degree of breakage as a function of energy input of a sample of ore using a drop weight apparatus. It is widely used in the mining industry to evaluate the ore's breakage characteristics and to optimize grinding circuit design. The test involves breaking a sample of ore of a specific size using predetermined energy levels and measuring the size distribution of the resulting fragments [2]. The test results, including the specific energy consumption, Ecs (kWh/t) and the t_{10} parameter (percentage passing 1/10th of their original particle size), are then used to calculate the ore's breakage parameters A and b values from Equation (1) [3]

$$t_{10} = A \times (1 - e^{-b \times Ecs}) \quad (1)$$

The standard test method requires about 50 kg of rock, so is only suited to bulk samples or requires the compositing of long lengths of precious drill core. Due to these sample constraints and the cost of the test, only a few samples are tested for an entire orebody in the design phase and only occasionally are samples tested during production. To help overcome these limitations, the standard testing method was modified by Morrell [4] to enable the use of a reduced mass of a sample and thus the application of the DWT to drill core samples. The resultant SMC (Steve Morrell comminution) method has enabled the DWT outcomes to be applied to a far greater extent across drill cores and extended into orebody characterisation. Morrell [4] also introduced new energy-size reduction equations based on the outputs of the SMC test, demonstrating its validity across several comminution processes. It does, however, remain somewhat costly and uses 20 kg of core (representing about 15 m of split core). The Geopyörä test aimed to further reduce the uptake barrier to testing more of the drill core within an orebody, thus providing greater resolution to modelling and predicting the mill performance and mine production over the life of a mine.

The purpose of this paper is to assess the accuracy and precision of the Geopyörä breakage test (GPT) energy and force measurements in order to provide the levels of confidence required by the industry for uptake of this ore characterisation technique. The accuracy was indirectly assessed by comparing the fragmentation (t_{10} parameter) obtained when testing duplicate samples at the same energy level using the well-established drop weight test (DWT). In addition, the precision (i.e., reproducibility) of the test was evaluated using a Round-Robin methodology [5], similarly to that reported by Mosher [6] and Weier [7] in comparative testing undertaken by over 30 metallurgical laboratories located around the world to determine the precision of the Bond test [8].

1.1. Equipment Description

The testing method is described in some detail in Bueno et al. [1], so only an overview of the pertinent operating principles is provided here. The principle of the method is illustrated in Figure 2, with the idealised force response during a breakage event and the change in angular velocity of the wheel during a breakage event. The counter-rotating wheels allow the automated feeding of rocks one at a time through the spinning wheels, with no requirements for stopping, resetting, and sweeping away broken fragments between each rock breakage.

The mechanical set-up enables measurement of the applied force during each breakage event with a loadcell providing a high sampling frequency of 5000 Hz to provide a force-to-fracture value, as illustrated in Figure 3. The force applied to break a rock with a given degree of compression is a function of the rock compressive strength. The force plot can thus be related to standard rock strength measures, an aspect not explored in this paper. Once the force measured by the load cell, due to rock fracture between the wheels, surpasses a predetermined threshold (Ft), the recording of force will commence for a specific duration (t_1–t_0). The highest force peak within this timeframe (Fp) will be registered.

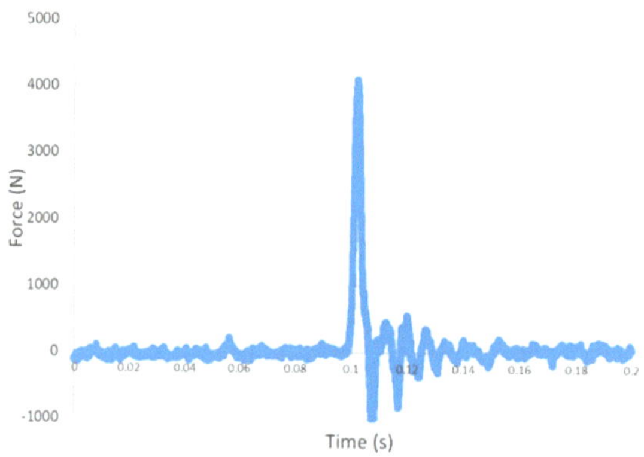

Figure 3. Example of a force plot during rock fracture.

The traditional ore characterisation of the drop weight test (DWT) [2,4] controls the energy input through the kinetic energy of a falling weight. Regardless of what is absorbed during the fracture event, this is given as the input energy to the breakage event. From this the specific comminution energy (Ecs in J/g or kWh/t) is calculated. In contrast, for the Geopyörä, the crushing is conducted at a stiff (fixed) gap and the energy is measured through the loss of momentum of the spinning wheels [1]. Thus, the breakage energy used for characterisation is not a controlled input, but rather a measure of the response of the system to the breakage characteristics of the rock—determined by the applied stress and properties intrinsic to the rock material (mineral composition, texture, grain structure, internal flaws, microcracks, etc.). By varying the degree of reduction, i.e., the ratio of the crushing gap to the particle diameter, a range of breakage energies can be achieved to map the response of the rock. It should be noted that the use of fixed energy points on the typical Ecs vs. t_{10} plot, which are then translated to the breakage function parameters of A and b, is only a matter of standard testing procedure.

The Geopyörä distinguishes itself from traditional drop weight tests by its ability to measure the breakage force and energy on a per-particle basis within a given sample. This unique feature enables the generation of distinct distributions of breakage energy and force for each sample, as depicted in Figures 4 and 5. The detailed nature of these distributions facilitates the application of statistical analyses, including the Student's *t*-test and ANOVA, providing a robust framework for comparing outcomes across varying samples.

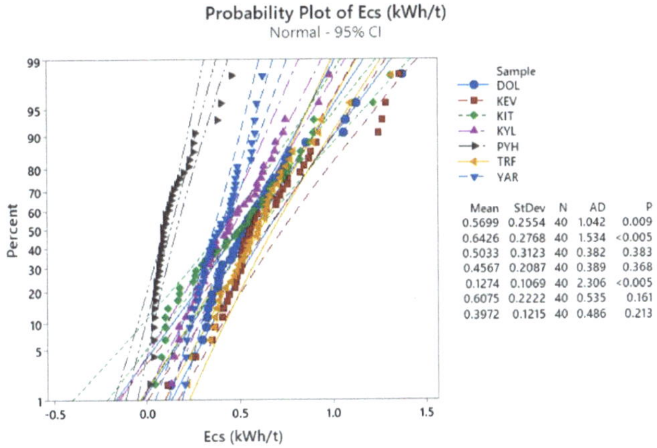

Figure 4. Geopyörä Ecs probability distribution.

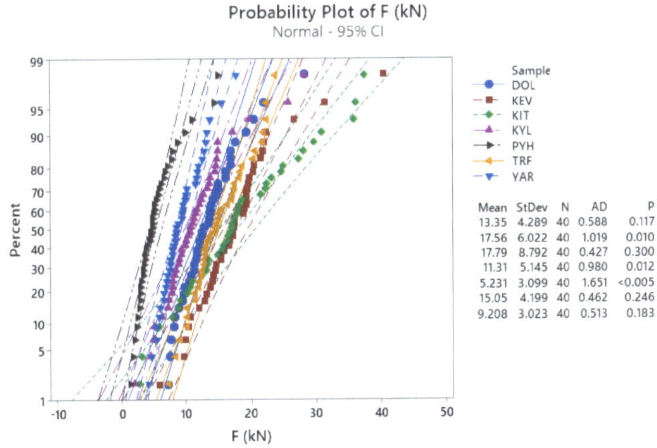

Figure 5. Geopyörä force probability distribution.

The energy consumed in each breakage event is measured via the momentum loss of the crushing wheels. Triggered by the release of a rock from the feed system, the power is disconnected from the direct-drive system and the breakage wheels are allowed to be idle. The resultant momentum loss arising from the rock breakage provides the measure of energy applied to compressing and breaking the rock particle. The losses due to mechanical back electromotive force and friction are carefully calibrated for every machine and accounted for in the energy calculation [1].

The mechanical operation of the Geopyörä is described in more detail in [1]. In such a design it is necessary to measure the absorbed energy per rock breakage with sufficient precision, while ensuring non-slip grip and compression of the rocks to the point of fracture.

1.2. Assessing Accuracy

In wishing to assess the accuracy of a new testing device, there are two major aspects to be verified:

Precision: measurements are reproducible and replicable between different test devices.

Accuracy: the measurement yields a value within a required uncertainty when assessed against an absolute value, such as in the SI standards.

However, there is no 'truth' of the breakage energy or rock strength in terms of the energy needed to produce crushed product. This has been explored by a number of researchers trying to define the efficiency of comminution devices. Fuerstenau [9] assessed the absolute efficiency of breakage energy to be below 1%, while Tromas [10] concluded the maximum possible efficiency is in the range of 2.5% to 7.5%. All devices have an inherent inefficiency. One of the best benchmarks for measuring breakage as a function of the energy input is the instrumented loadcell method, in which the energy absorbed by the rock is measured, as opposed to the applied energy. The impact loadcell studies of Tavers and King [11] and of Bourgeois and Banini [12] provide a measure of the absorbed versus applied energy, with measures varying depending on the specific applied energy and the rock competence—whereby softer rocks absorb less energy than stiffer rocks. Tavares [13] measured the absorbed energy as varying between 67% and 99% of the applied impact energy. The objective function for determining the accuracy of a new test thus tends to be taken relative to existing tests. The objective at this stage of introducing the new test is to replicate the measures of the DWT, so as to enable these well-accepted measures to be propagated across the orebody in far greater detail than can be achieved using DWT equipment.

Accuracy: statistically match the measurements derived from full DWTs.

A complication of comparing to another test is that the base test has an inherent degree of precision. Thus, the precision of the DWT should be included in an assessment of accuracy. The work of Tavares and King [11] illustrates the natural variability of precision measures conducted on the impact load cell (UFLC) apparatus, which is in essence an instrumented DWT. JK Tech has conducted a study of reproducibility between repeat tests conducted on the JK DWT [7]. The study revealed that when the same rock sample was tested at various laboratories worldwide, the results from the Bond test differed by ±14.2%, while for the SMC test, the difference was ±9.2%. This uncertainty must be incorporated into any test of accuracy.

2. Materials and Methods

The overall approach taken in this work to assess precision and accuracy is through comparative data and repeat tests using carefully sampled ore samples.

2.1. Accuracy Assessment Experiments

Samples from seven different ore deposits with different mineralogy, as is shown in Table 1, were used in this research. Bulk samples were crushed and sieved to obtain particles of 22 × 19 mm. From each sample population, eight samples of 20 particles were selected. To ensure consistency between duplicate samples, the particle selection methodology of the SMC Test® [4] was applied, which limits the particle mass to within the mean ±30%.

Table 1. Samples used in this test work.

Sample	Ore Type
DOL	Calcium carbonate
KEV	Copper
KIT	Gold
KYL	Copper-zinc
PYH	Copper-zinc
TRF	Nickel-cobalt
YAR	Phosphate

To assess the accuracy of energy measurement in the Geopyörä test, various tests were conducted. In total, 56 sets of 20 particles were tested, with each ore type having four sets tested using the Geopyörä and four sets subjected to a custom drop weight test at the University of Sao Paulo (USP). This resulted in eight sets for each ore type, enabling a detailed analysis of the results.

The Geopyörä tests were conducted with two repeats (1 and 2) performed at a wider gap setting of 50% of the particle size, and two other repeats (3 and 4) carried out at a closer gap setting of 25%. The gap ratio is defined as the proportion between the geometrical means of the tested particles, which was 20.6 mm in this case, and the gap aperture between the wheels. The gap ratios of 50% and 25% represent low and high specific energy levels, respectively.

The standard drop weight test uses fixed, predetermined energy values, whereas the Geopyörä test measures the actual specific breakage energy of each particle, which is a response of the tested material rather than a test input. As a result, custom drop weight tests were conducted by adjusting the energy input to match the mean specific energies as measured by the Geopyörä test, ensuring that the results were directly comparable to those obtained from the Geopyörä test.

The products of both the Geopyörä and drop weight tests were sieved to determine the t_{10} parameter, which was compared. The hypothesis is that the Geopyörä energy measurements can be quantified by comparing the t_{10} at equivalent Ecs values to the DWT. This methodology had been previously attempted by Chaves Matus [14], but he had to use Equation (1) to interpolate t_{10} values at the standard Ecs values used in the JK DWT.

2.2. Precision Assessment (Round-Robin)

Round-robin tests, also known as interlaboratory tests or proficiency tests, are a type of testing where a set of samples are distributed to multiple laboratories or testing facilities to determine their performance and compare results. The samples used in the tests are typically prepared in such a way as to represent a range of materials and/or properties that may be encountered in real-world applications [5].

Each laboratory conducts the same test on the samples and reports their results. The results are then compiled and analysed to determine the degree of variability between the different labs and the accuracy and precision of the testing methods. The data obtained from the round-robin tests can be used to identify sources of error and variability in testing methods and to improve the quality and consistency of the testing process.

Round-robin tests were conducted with a few duplicate samples from the bulk ore samples that had previously been used in the accuracy assessment—except for the PHY ore. The repeat tests were conducted using four different Geopyörä devices/laboratories (anonymised as A, B, C and D) under the same operating conditions (i.e., particle size, gap aperture and wheel speed) for each ore type, which are summarized in Table 2. The precision of both energy and force measurements were assessed using analysis of variance (ANOVA) tests.

Table 2. Comparative test samples, devices and operating conditions.

Sample	Particle Size (mm)	Gap (mm)	Speed (rpm)	Number of Repeat Tests per Sample/Device				
				A	B	C	D	Total
DOL	−22.4 + 19	10.3	80	2	2	2	2	8
KEV	−22.4 + 19	10.3	80	2	2	2	2	8
KIT	−16 + 13.2	7.3	80	2	2	1	-	5
KYL	−16 + 13.2	7.3	80	2	2	1	-	5
TRF	−16 + 13.2	7.3	80	2	2	1	-	5
YAR	−19 + 16	8.7	80	2	2	1	-	5
Total number of tests per device				12	12	8	4	36

3. Results and Discussion

3.1. Accuracy Assessment

Before sending the duplicate samples for the comparative custom drop weight tests, the precision of the specific energy and force measurements made using the Geopyörä were statistically confirmed using the Student's t-test on the repeat tests. The mean Ecs and force values with a confidence interval of 95% for both low energy (T1 and T2) and high energy (T3 and T4) repeat tests are plotted as repeat pairs in Figure 6. The error bars show 95% confidence limits, allowing a visual assessment of the repeat data for each test overlap between repeats—which is obeyed for all repeat tests for both Ecs and force. It can be seen that the tests with a smaller gap (3 and 4) absorb far higher energy, shown in the graphs on the right, than the ones with a wider gap (1 and 2), typically around twice the specific energy. The peak fracture force, shown in the lower plots, does increase with the higher compression of tests 3 and 4, but by a lesser degree than the energy.

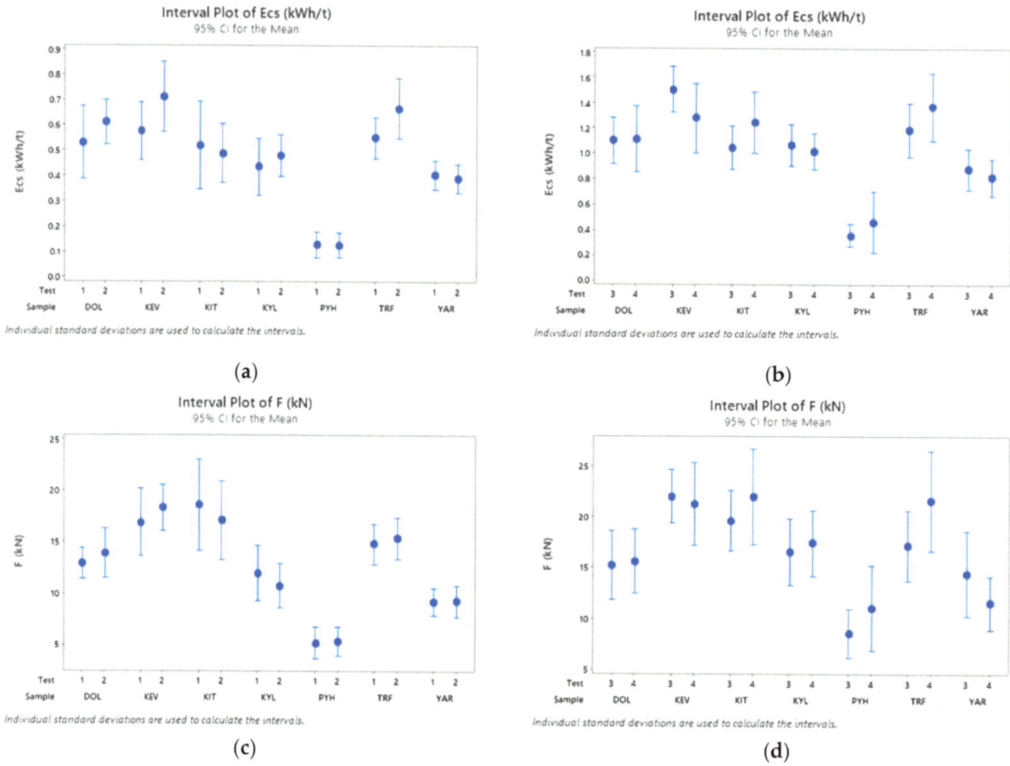

Figure 6. Mean Ecs and force with 95% confidence intervals for low energy tests, T1 (**a**) and T2 (**b**) in the first row and high energy tests, T3 (**c**) and T4 (**d**) in the second row, respectively.

The t-test values are presented in Table 3. If the p-values are less than 0.05, we could reject the null hypothesis that there is a difference between the means and conclude that a significant difference does exist. However, since all the p-values were larger than 0.05, we cannot conclude that a significant difference exists with 95% confidence between pairs of repeat tests.

Table 3. *t*-test *p*-values.

Sample	*t*-Test *p*-values			
	T1 and T2		T3 and T4	
	Ecs	Force	Ecs	Force
DOL	0.327	0.474	0.926	0.857
KEV	0.118	0.456	0.160	0.742
KIT	0.749	0.598	0.162	0.380
KYL	0.510	0.453	0.636	0.705
PYH	0.933	0.876	0.409	0.292
TRF	0.099	0.668	0.270	0.132
YAR	0.706	0.952	0.541	0.235

The t_{10} fragmentation parameter from the products of the Geopyörä and drop weight tests conducted at the same energy levels are presented in Table 4. No significant difference was observed and the t_{10} values were within ±15% of each other, as show in Figure 7. This indicates that the Geopyörä test energy measurements were accurate and reliable in estimating the breakage characteristics of the tested rock material. Therefore, the hypothesis that the Geopyörä energy measurements are accurate if we obtain the same fragmentation that is obtained when we break the same rock material in a drop weight test using the same energy input can be confirmed.

Table 4. Mean specific energy, Ecs (kWh/t), and t_{10} values for Geopyörä (GPT) and drop weight tests (DWT).

Sample	Test 1				Test 2				Test 3				Test 4			
	Ecs	t_{10} (%)			Ecs	t_{10} (%)			Ecs	t_{10} (%)			Ecs	t_{10} (%)		
		GPT	DWT	% diff		GPT	DWT	% diff		GPT	DWT	% diff		GPT	DWT	% diff
DOL	0.51	11.1	12.2	9%	0.62	14.4	16.2	11%	1.09	28.9	30.2	4%	0.98	26.3	26.7	1%
KEV	0.60	13.9	13.2	−5%	0.71	16.0	15.8	−1%	0.91	32.4	31.6	−3%	1.22	27.6	30.0	8%
KIT	0.50	11.5	10.2	−13%	0.49	9.8	9.6	−2%	1.06	23.0	24.3	5%	1.35	29.4	31.1	5%
KYL	0.41	14.2	13.4	−6%	0.47	15.6	17.6	11%	1.06	38.1	39.5	4%	0.98	36.8	36.8	0%
PYH	0.12	26.7	27.0	1%	0.13	26.7	29.3	9%	0.35	54.2	52.8	−3%	0.4	53.3	52.2	−2%
TRF	0.55	11.4	9.8	−16%	0.67	12.3	14.3	14%	1.25	27.9	25.8	−8%	1.29	27.5	28.1	2%
YAR	0.42	19.9	19.5	−2%	0.40	19.9	21.0	5%	0.89	42.9	41.5	−3%	0.81	38.6	37.2	−4%

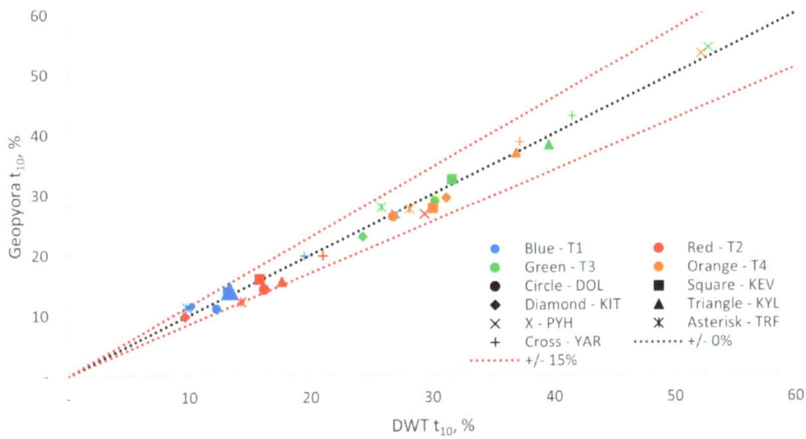

Figure 7. Parity chart of t_{10} parameter measured in the Geopyörä vs. DWT.

The relationships between Ecs and t_{10}, which are plotted in Figure 8, were also assessed to identify potential differences in the fitting of breakage parameters A and b from Equation (1). Since the Geopyörä and the drop weight tests were conducted at the same energies and the measured fragmentation (t_{10}) was practically the same, no significant difference was found in the resulting A and b parameters obtained using the Geopyörä and drop weight data.

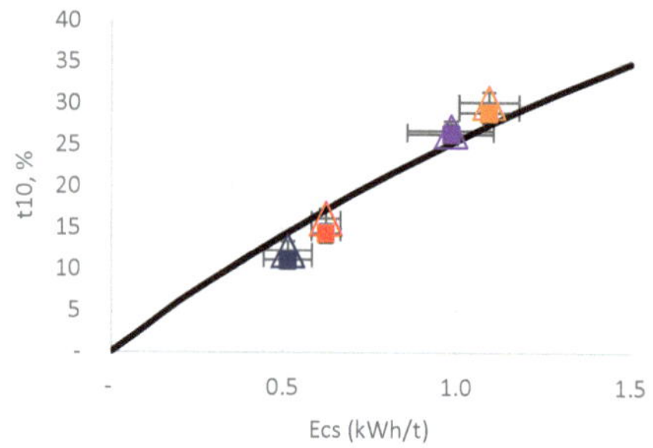

(**a**) DOL Axb fit

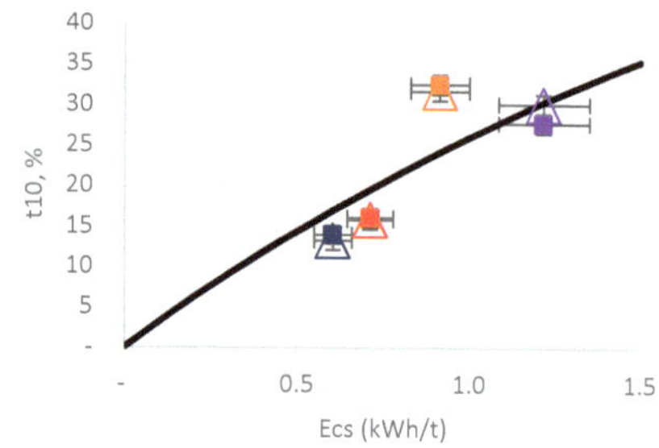

(**b**) KEV Axb fit

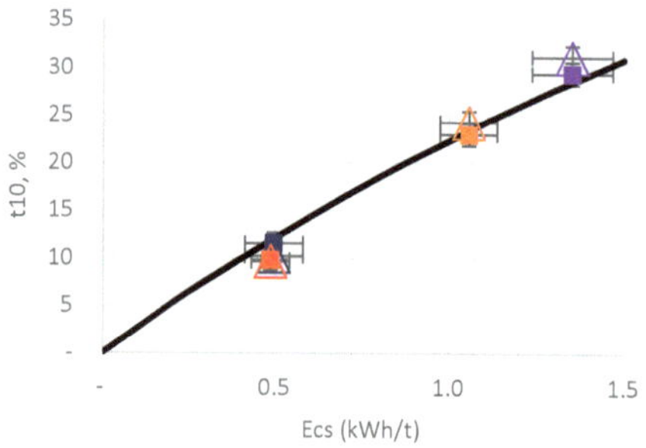

(**c**) KIT Axb fit

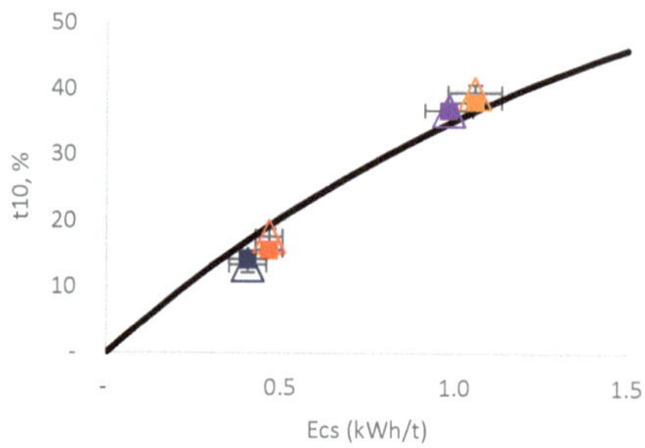

(**d**) KYL Axb model fit

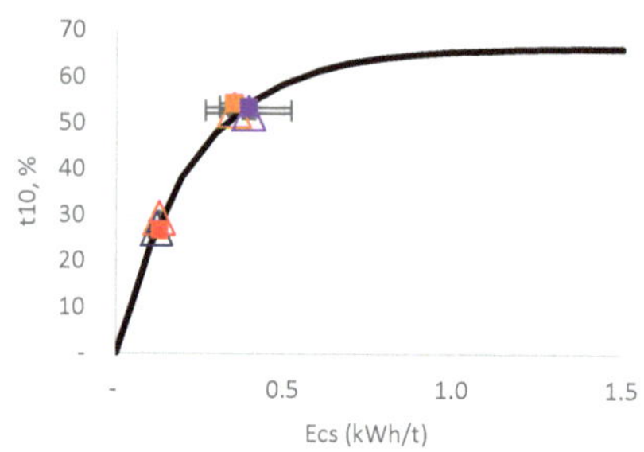

(**e**) PYH Axb fit

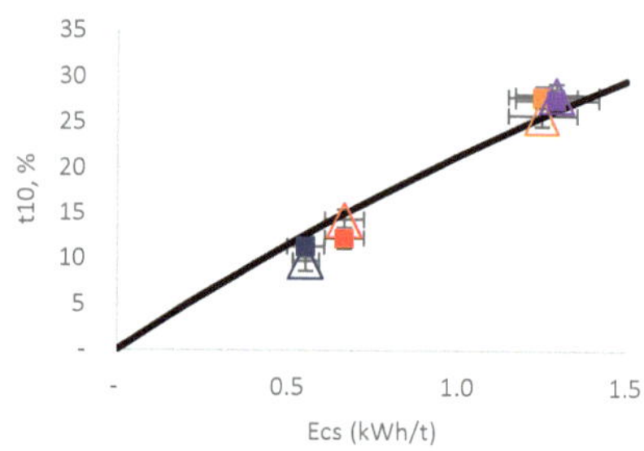

(**f**) TRF Axb fit

Figure 8. *Cont.*

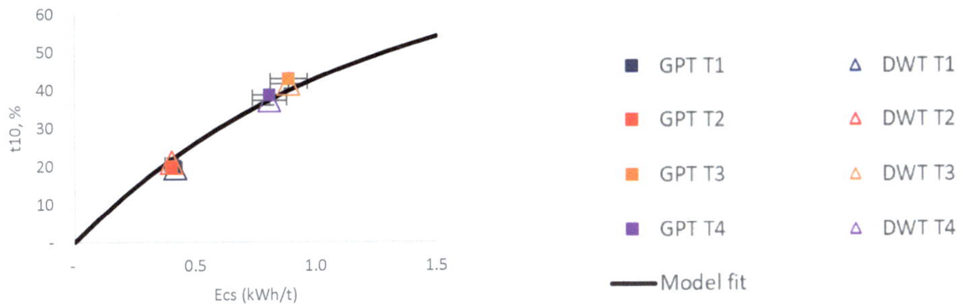

(g) YAR Axb fit

Figure 8. t_{10} and Ecs plots for all seven samples.

As presented in Table 5, the results were within −2.7 to 6.1% from one another. While this is a confirmation that the Geopyörä test can be a reliable alternative to the DWT in estimating the breakage characteristics of rock material, a more extensive validation is presented in Bueno et al. [15].

Table 5. Comparison of the fitted Axb parameters using the Geopyörä and DWT test data.

Sample	Fitted A and b		Difference	
	Geopyörä	DWT	Value	%
DOL	39.8	42.4	2.6	6.1%
KEV	41.6	42.2	0.6	1.4%
KIT	34.2	35.6	1.4	3.9%
KYL	61.7	63.8	2.1	3.3%
PYH	323	326	3	0.9%
TRF	33.3	32.8	−0.5	−1.5%
YAR	75	73	−2	−2.7%

3.2. Precision Test Results

The Geopyörä test measures the specific comminution energy, Ecs (kWh/t), and applied force, F (kN), for each single particle in a sample, resulting in distributions of Ecs and F values. The data obtained for the repeat tests conducted with six different ore samples across four devices are presented as box plots in Figures 9 and 10. The range of Ecs or force in any given test represents the natural variability of the ore, within a carefully constructed subsample of near identical rocks, and the difference in rock dimensions within a screened size range. It should be noted that such information is not available for the DWT results as no input data are measured.

A one-way ANOVA test was conducted to compare the mean Ecs and F values of each test to determine whether they were significantly different or not. The calculated p-values presented in Table 6 are greater than 0.05, confirming the precision of the tests as there are no significant differences with 95% confidence.

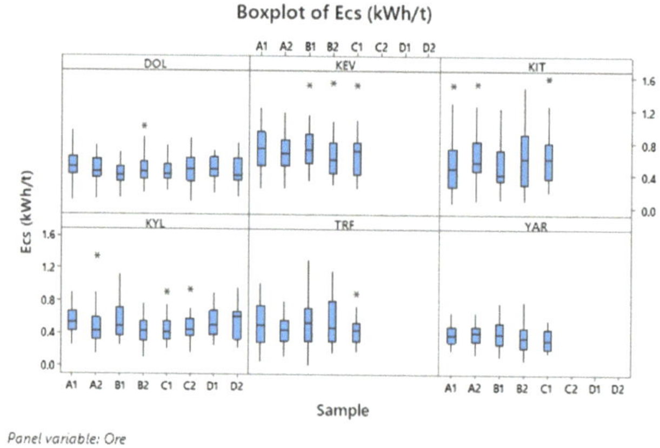

Figure 9. Repeat specific energy measurements.

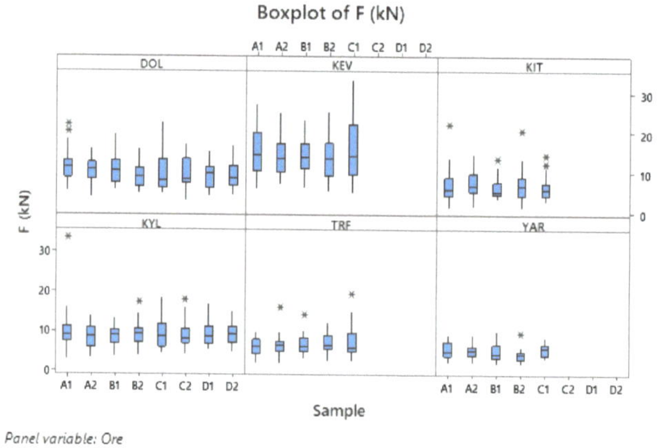

Figure 10. Repeat force measurements.

Table 6. ANOVA calculated *p*-values.

Sample	ANOVA *p*-Values	
	Ecs (kWh/t)	F (N)
DOL	0.354	0.065
KEV	0.821	0.843
KIT	0.516	0.149
KYL	0.136	0.275
TRF	0.258	0.552
YAR	0.393	0.369

Since there are no significant differences among the tests, the overall precision of the Ecs and orce measurements was determined on the basis of the 95% confidence interval of the mean calculated using all the available data for each ore type. The summary results presented in Table 7 show an average precision of 6.8% for specific energy measurements and 6.3% for force measurements across all different devices and ore types.

Table 7. Mean, standard error of the mean, standard deviation and 95% confidence intervals for Ecs and force measurements.

Ore	N	Ecs (kWh/t)			95% CI		F (kN)			95% CI	
		Mean	SE	SD	(+/−)	%	Mean	SE	SD	(+/−)	%
DOL	199	0.46	0.01	0.17	0.02	5.1%	10.1	0.3	3.7	0.51	5.1%
KEV	137	0.69	0.02	0.27	0.05	6.5%	14.5	0.5	5.4	0.90	6.2%
KIT	146	0.59	0.03	0.33	0.05	9.1%	6.8	0.3	3.4	0.55	8.1%
KYL	198	0.50	0.01	0.20	0.03	5.6%	9.2	0.2	3.5	0.49	5.3%
TRF	143	0.50	0.02	0.25	0.04	8.3%	6.8	0.2	2.8	0.46	6.8%
YAR	140	0.38	0.01	0.15	0.02	6.3%	5.0	0.2	1.9	0.31	6.2%
					Mean	6.8%				Mean	6.3%

4. Conclusions

In conclusion, this study evaluated the accuracy and precision of the Geopyörä breakage test for assessing the specific energy required for rock fragmentation. The results showed that the Geopyörä test provides accurate energy measurements, as confirmed by the close agreement with the results obtained from the custom drop weight tests. The precision of the Geopyörä test was also demonstrated through round-robin tests, which showed good agreement of the breakage energy and force measurements conducted with several ores and four different devices. The repeatability achieved by the Geopyörä test outweigh that achieved with the commercial drop weight test and the Bond test. Therefore, the Geopyörä test is a reliable and efficient method for accurately measuring the specific breakage energy of rock breakage, with the potential to improve the efficiency and sustainability of mining operations. Further research is needed to investigate the accuracy of its force measurements against other methods, such as the UFLC [11]. More extensive round-robin tests with other rock types and across more laboratories will also be conducted in the near future.

Author Contributions: M.B.: Conceptualization, Data curation, Formal analysis, Investigation, Methodology, Resources, Project administration, Writing—original draft, Writing—review and editing, and Supervision. T.A.: Formal analysis, Data Curation, Validation, Visualization, Writing—review and editing, and Supervision. L.L.: Formal analysis, Investigation, Data Curation, Validation, Visualization, Writing—review and editing, and Supervision. M.P.: Writing—review and editing and Supervision. H.D.: Resources. All authors have read and agreed to the published version of the manuscript.

Funding: This research received no external funding.

Data Availability Statement: The data presented in this study are available on request from the corresponding author due to privacy reasons.

Conflicts of Interest: Marcos Bueno is employed by the Geopyörä Oy company. The remaining authors declare that the research was conducted in the absence of any commercial or financial relationships that could be construed as a potential conflict of interest.

References

1. Bueno, M.; Torvela, J.; Chandramohan, R.; Matus, T.C.; Liedes, T.; Powell, M.S. The double wheel breakage test. *Miner. Eng.* **2021**, *168*, 106905. [CrossRef]
2. Napier-Munn, T.; Morrell, S.; Morrison, R.; Kojovic, T. *Mineral Comminution Circuits: Their Operation and Optimisation*, 1st ed.; Julius Kruttschnitt Mineral Research Centre: Indooroopilly, Australia, 1996.
3. Narayanan, S.S.; Whiten, W.J. Determination of comminution characteristics from single-particle breakage tests and its application to ball-mill scale-up. *Trans. Inst. Min. Metall. Sect. C* **1988**, *97*, C115–C124.
4. Morrell, S. Predicting the specific energy of autogenous and semi-autogenous mills from small diameter drill core samples. *Miner. Eng.* **2004**, *17*, 447–451. [CrossRef]
5. Napier-Munn, T.J. *Statistical Methods for Mineral Engineers: How to Design Experiments and Analyse Data*; Julius Kruttschnitt Mineral Research Centre, University of Queensland: Indooroopilly, Australia, 2014.

6. Mosher, J.B.; Tague, C.B. Conduct and precision of Bond grindability testing. *Miner. Eng.* **2001**, *14*, 1187–1197. [CrossRef]
7. Weier, M.L.; Chenje, T. Accuracy of the Bond ball mill test and its implications. In Proceedings of the 14th International Mineral Processing Conference & 5th International Seminar on GeoMetallurgy, Santiago, Chile, 28–30 November 2018; Procemin-Geomet: Santiago, Chile, 2018.
8. Bond, F.C. Crushing and grinding calculations. *Br. Chem. Eng.* **1961**, *6*, 378–385.
9. Fuerstenau, D.W.; Abouzeid, A.Z.M. The energy efficiency of ball milling in comminution. *Int. J. Miner. Process.* **2002**, *67*, 161–185. [CrossRef]
10. Tromas, D. Mineral comminution: Energy efficiency considerations. *Miner. Eng.* **2008**, *21*, 613–620. [CrossRef]
11. Tavares, L.M.; King, R.P. Single-particle fracture under impact loading. *Int. J. Miner. Process.* **1998**, *54*, 1–28. [CrossRef]
12. Bourgeois, F.; Banini, G. A portable load cell for in-situ ore impact breakage testing. *Int. J. Miner. Process.* **2002**, *65*, 31–54. [CrossRef]
13. Tavares, L.M. Breakage of Single Particles: Quasi-Static. In *Handbook of Powder Technology*; CRC Press: Boca Raton, FL, USA, 2007; Volume 12. [CrossRef]
14. Chavez Matus, T. The Geopyörä breakage test for geometallurgy. In Proceedings of the IMPC 2020: XXX International Mineral Processing Congress, Cape Town, South Africa, 18–22 October 2020.
15. Bueno, M.; Almeida, T.; Powell, M.S. Extensive validation of a new rock breakage test. In Proceedings of the SME Annual Meeting, Denver, CO, USA, 26 February–1 March 2023.

minerals

MDPI

Article

Enhancing Iron Ore Grindability through Hybrid Thermal-Mechanical Pretreatment

Sefiu O. Adewuyi [1,2,*], Hussin A. M. Ahmed [2], Angelina Anani [1], Abdu Saeed [3,4], Haitham M. Ahmed [2], Reem Alwafi [3] and Kray Luxbacher [1]

1 Department of Mining and Geological Engineering, University of Arizona, Tucson, AZ 85721, USA; angelinaanani@arizona.edu (A.A.); kraylux@arizona.edu (K.L.)

2 Mining Engineering Department, King Abdulaziz University, Jeddah 21589, Saudi Arabia; hussien135@gmail.com (H.A.M.A.); hmahmed@kau.edu.sa (H.M.A.)

3 Department of Physics, King Abdulaziz University, Jeddah 21589, Saudi Arabia; abdusaeed79@hotmail.com (A.S.); ralwafy@kau.edu.sa (R.A.)

4 Department of Physics, Thamar University, Thamar 87246, Yemen

* Correspondence: sadewuyi@arizona.edu or sefiuadewuyi@gmail.com

Abstract: Grinding is an important process of ore beneficiation that consumes a significant amount of energy. Pretreating ore before grinding has been proposed to improve ore grindability, reduce comminution energy, and enhance downstream operations. This paper investigates hybrid thermal mechanical pretreatment to improve iron ore grinding behavior. Thermal pretreatment was performed using conventional and microwave approaches, while mechanical pretreatment was conducted with a pressure device using a piston die. Results indicate that conventional (heating rate: 10 °C; maximum temperature: 400 °C), microwave (2.45 GHz, 1.7 kW, 60 s), and mechanical (14.86 MPa, zero delay time) pretreatments improved the studied iron ore grindability by 4.6, 19.8, and 15.4%, respectively. Meanwhile, conventional-mechanical and microwave-mechanical pretreatments enhanced the studied iron ore grindability by 19.2% and 22.6%, respectively. These results suggest that stand-alone mechanical pretreatment or microwave pretreatment may be more beneficial in improving the grinding behavior of the studied fine-grain iron ore sample. The results of the mechanical pretreatment obtained in this study may be used in a simulation of the HPGR system for grinding operations of similar iron ore

Keywords: microwave heating; thermal treatment; comminution energy; ore grinding; work index

1. Introduction

Huge energy consumption for ore grinding [1] and tailing management [2,3] are major challenges in mineral processing. Reducing energy consumption and sustainable tailing management are among the current challenges in the mining industry. Efficient comminution operation improves mineral liberation, leading to improved mineral recovery and low tailing production, making comminution a critical operation in mineral processing [4]. Therefore, efforts to improve comminution operations are receiving significant attention. With ore grades declining globally, demand for metals and industrial minerals increasing, and the transition to industrial 4.0, the energy requirement for comminution processes is increasing, causing greater concern than ever before. Among the mining operations, grinding consumes up to 70% of the total energy consumption [5,6]. Studies have been performed to reduce grinding energy by focusing on factors to improve grinding performance. Among the factors affecting grinding performance are grinding media, machine type, operating conditions, and ore characteristics [7–9]. Attempts have been made to improve ore grinding operations by optimizing the ore-to-media ratio, mill-length-to-diameter ratio, mill speed, media size distribution, and varying grinding balls such as mild steel, stainless steel, and nano-ceramic [5]. Furthermore, efforts have been expended on ore comminution machine design, resulting in the production of the high-pressure grinding roll (HPGR), which can minimize energy loss as noise and heat—leading to improved ore grinding and reduction in comminution energy [10]. However, HPGR produces a coarser particle size distribution due to the edge effect [11]. Since the grindability of ore is a material's characteristic, researchers proposed that ores may be pretreated to improve their physical properties, which may enhance grinding behavior and lead to significant energy savings [5]. Different ore pretreatment methods have been explicitly presented in the literature [5]. Thermal pretreatment is among the focused techniques that may be performed by conventional (via furnace) or microwave (MW). Both heating approaches have been employed for several minerals with varying degrees of heating responses and induced cracks on the tested ore/mineral samples [5]. The type of cracks developed by the material samples and the extent of cracks determine the effect of absorbed heat energy on the material's grinding behavior [5]. Ratan et al. [12] discussed that conventional heating of hematite ore to 400 °C followed by water quenching may reduce grinding energy by 45%. Omran et al. [13] investigated the effect of conventional heating on the grinding behavior of phosphorus-oolitic iron ore and results indicated that particles passing a sieve size of 0.125 mm increased from 46.6 to 50.8% after heating to 600 °C for 1 h. Based on the available literature, little is known about the grinding behavior of conventionally heated iron ore, but the approach has been widely studied to convert hematite to magnetite [14]. Most researchers have focused on the grinding behavior of conventionally heated quartzite and calcite [15].

For microwave heating, studies are increasingly focusing on reducing grinding energy requirements in the mining industry through microwave pretreatment of ores. Walkiewicz et al. [6] found that cracks developed on iron ore samples after being subjected to microwave heating, resulting in an improvement in work index up to 23.7%, for hematite obtained from Republic mine, Michigan, USA. Micro-fractures were observed on microwave-treated hematite selected from Orissa, India, that resulted in an improvement in the specific rate of breakage up to 50% [16]. Song et al. [17] reported that intergranular cracks developed on hematite ore after microwave irradiation, which improved hematite liberation from quartz and apatite by up to 30%. Singh et al. [18] pretreated iron ore (containing hematite and jasper) at 900 W for 5 min. Grindability test results indicated that microwave pretreatment decreased d_{80} by 20.83% compared to untreated ore samples [18]. Abdur Rasyid et al. [19] studied the power-saving capability of microwave pretreatment during the crushing operation of kimberlite rock using a single-roll crusher. It was found that specific crushing energy of up to 18% can be saved using microwave pretreatment. Hao et al. [20] studied the microwave-damage mechanism of magnetite ore. Their results showed that increasing microwave power led to deceased ore mechanical properties due to intergranular and intragranular cracks developed after microwave irradiation, leading to improved crushing degree. Omran et al. [13] compared the effect of conventional and microwave pretreatments and found that intergranular fractures developed between hematite and its gangue (fluoroapatite and chamosite) after microwave treatment, while a small proportion of micro-cracks were noticed on the ore's surface after conventional heating. Grindability tests showed that the weight percentage passing sieve size 0.125 mm increased by 13.16% after microwave pretreatment compared to a 4.2% increase for conventional pretreatment [13]. Recently, results of possible energy reduction due to microwave pretreatment of different ores/minerals have been presented in the literature [5]. Findings indicated that despite intensive research on microwave applications for improved ore grinding for better energy utilization in the mining industry, there is little research on iron ore (Table 1).

The piston-die test is usually employed to study particle-bed breakage for understanding interparticle breakage, estimating the energy required for comminution by compressive load, and predicting the breakage behavior of HPGR [21,22]. A comparison of the particle liberation using particle-bed breakage (by piston-die compression), hammer mill, and ball mill has been discussed in the literature [23]. Particle-bed breakage generally enhanced the particle liberation for clinker [24], calcite [25], and quartz [22]. Meanwhile, low-grade porphyry copper ore at a size fraction below 150 μm exhibited similar particle liberation in hammer mill and piston-die compression [23]. The selected iron ore samples in this study consist of interwoven fine-grain hematite, magnetite, and diopside; these usually require fine grinding, consume a high amount of comminution energy, and cause metallurgical challenges [26]. Pretreating this type of iron ore may improve its grinding behavior. This paper investigates hybrid thermal-mechanical pretreatments to enhance the selected iron ore grinding behavior. Thermal pretreatment was studied using conventional and microwave approaches, while mechanical pretreatment was conducted using a piston-die test. The grinding behavior of the investigated pretreatment approaches was evaluated using a laboratory standard Bond ball mill.

Table 1. Microwave pretreatment of iron ore samples for improved ore grindability (MW frequency: 2450 MHz), temp. = temperature.

Sample Location	Mineral Phases	Size Fraction (mm)	Sample Mass (g)	MW Power (kW)	MW Time (min)	Average MW Temp. (°C)	Improvement in Grindability (%)	Reference
Republic mine, Michigan	Hematite, quartz	−3.35	350	3.0	3.5	840	23.7	[6]
Empire mine, Michigan	Magnetite, quartz	−3.35	350	3.0	3.5	840	21.4	[6]
Orissa, India	Hematite, alumina, silica	−19.05 + 12.7	500	0.9	2.0	148	50	[16]
Hubei province, China	Hematite, quartz, chlorite, apatite	-	50	1.2	5.0	-	-	[17]
Aswan region, Egypt	Hematite, quartz, fluoroapatite, chamosite	-	100	0.9	1.0	546	-	[13]
Joda, Odisha, India	Hematite, jasper	−3.35	50	0.9	5.0	510	20.8	[18]

2. Materials and Methods

2.1. Sample Collection and Preparation

The iron ore sample was collected from the Wadi Sawawin area, Tabuk Province, Saudi Arabia. This sample was considered in this study due to the increasing demand for iron ore around the world and the high level of energy consumption during grinding operations. Also, iron ore is known for its high dielectric and electric properties, which are crucial for investigating rock's response to thermal treatment via microwave or furnace, making it suitable for this study. The as-received samples (20–25 cm) were preliminary crushed by a hand-held hammer and further primary and secondary crushing operations were conducted using a laboratory jaw crusher (BB300 Mangan Retsch, Retsch-Allee Haan, Germany) and a laboratory roll crusher (Sew-Eurodrive GmbH & Co. KG, Bruchsal, Germany), respectively. These operations continued until a 100% passing sieve size of 3.35 mm, required for the standard Bond ball mill grindability (BBMG) test, was achieved (Figure 1). A quartzite sample collected from the Al Masane Al Kobra (AMAK, Nejran, Saudi Arabia) mining company [15] was used in this study for the optimization of mechanical pretreatment. The conning and quartering method (Figure 2) and a mechanical riffle-splitter (KHD Humboldt Wedag AG, Colonia-Allee 3, Cologne, Germany) were employed to obtain representative sub-samples. The same sampling procedure was used for the studied quartzite and iron ore samples. In this approach, the crushed sample was thoroughly mixed and shaped into a cone and then quarterly subdivided till sub-samples of approximately 5 kg each were obtained. One of the sub-samples was subdivided using a mechanical riffle-splitter until a mass equivalent to 700 cm^3 (required as the first feed for the BBMG test) was obtained. The average mass of five repeated experiments for the quartzite sample was 1282.5 g with a standard deviation of 0.2160 [15]. This indicates that the employed sampling procedure produced representative samples. Specifically, to obtain a sample of 1282.5 g for a BBMG test, a 5 kg sub-sample was thoroughly mixed for homogeneity and divided into four—each with approximately 1250 g, using a mechanical riffle-splitter. One 1250 g sub-sample was split into four with approximately 312.5 g each. One of these sub-samples was also divided into four. One of the obtained sub-samples was then divided into two to get approximately 39 g sub-samples each. A 39 g sub-sample was thoroughly mixed with a 1250 g sub-sample in a pan and formed a cone. The pan was then placed on a digital chemical balance where a spatula was used to reduce the mass by 6.5 g from four-quarter sides of the cone. To perform the particle size analysis of the sub-samples, eight sieves with size range 3150 µm–75 µm were selected [15]. The representative sub-samples obtained using a laboratory mechanical riffle-splitter were dry-screened by a laboratory electric sieve shaker (AS 200, Retsch, Retsch-Allee Haan, Germany) by placing the sample on the top sieve (3150 µm) [15]. The cumulative percentage of weight passings was calculated and recorded based on the weight passing each sieve. Particle size was plotted against the cumulative %weight passing to determine the 80% passing size (F_{80}) required for the grindability tests [15]. The repeated BBMG tests performed by employing the above sub-sampling procedure using a quartzite sample showed variation within ±0.4% [15], which falls within the acceptable industrial standard of ±3.4% [16]. Therefore, the sub-sampling procedure outlined above was applied to the studied iron ore sample. However, the feed mass obtained for the BBMG test of the investigated iron ore was 1710.5 g—indicating a packing density of 2.4436 g/cm^3. To get this mass, a stepwise mechanical riffle-splitter operation produced 1718.5 g, which was thoroughly mixed and shaped like a cone in a sample pan. This mass was then reduced to the required mass (1710.5 g) for the BBMG test, using the same approach as quartzite.

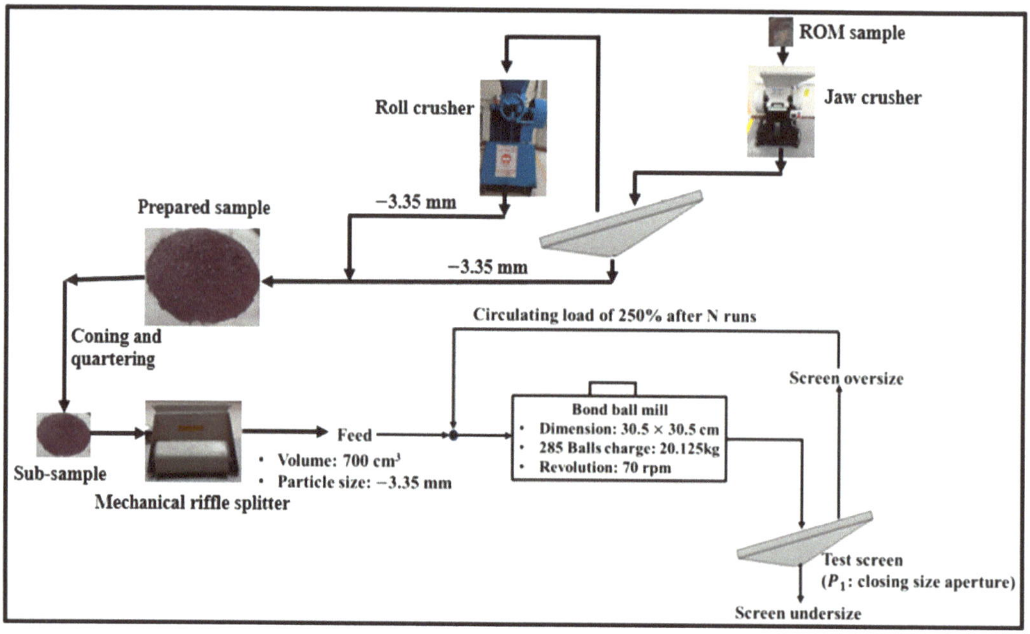

Figure 1. Sample preparation and Bond ball mill grindability (BBMG) test procedure (rpm—revolution per min, ROM—run-of-mine).

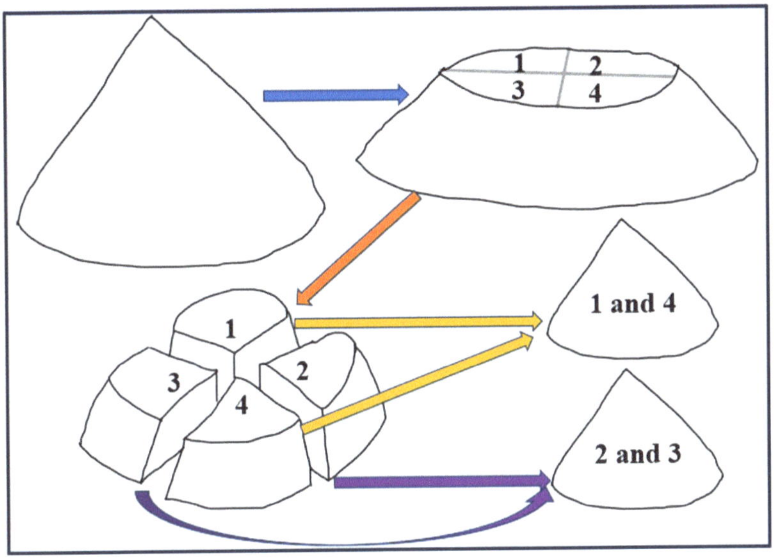

Figure 2. Coning and quartering technique (→: by relying on radial symmetry, a conical heap is divided into four, by cross; →: the sample is divided into four, along the cross; →: opposite quarters (1 and 4) are combined to form a sub-sample, →: opposite quarters (2 and 3) are combined to form a sub-sample).

2.2. Sample Characterizations

To perform chemical, mineralogy and morphology analyses of the studied samples, the laboratory mechanical riffle-splitter was used to appropriately obtain representative sub-samples from the prepared sample (Figure 1). Using a ball mill, the sample was ground to below 75 μm sieve size. The miniature sampling technique, using a spatula usually employed for powder material, was used to obtain sub-samples for chemical, mineralogy, and morphology analysis [15]. Mineralogy analysis was performed using the X-ray Diffraction method (Regaku, Ultima 1V X-ray diffractometer, Tokyo, Japan) as described in the literature [15]. Chemical analysis was also performed using a Fourier-transform infrared (FTIR) spectroscope (Nicolet iS50 FTIR, Thermo Fisher Scientific, Waltham, MA, USA) between 4000–400 cm^{-1} in transmittance mode. Morphology analysis was conducted using a Field Emission Scanning Electron Microscope (FESEM, JSM-7600F, JEOL Ltd., Musashino, Akishima, Tokyo, Japan) [15,27]. The description of FESEM for powder sample (without coating) characterization was adopted as presented in the literature [15,28,29]. The electric and dielectric properties of the studied samples were characterized to understand the selected sample's response to temperature. To do so, a cuboid-shaped iron ore sample was prepared with dimensions of 12, 10, and 2.5 mm. Sandpaper was used to smoothen the surfaces of the prepared sample, and the surfaces were then painted using a conductive platinum paste (Nanoshel LLC, Willmington, DE, USA) to form electrodes [30]. The iron ore cuboid was oven-dried at 150°C for 2 h. The electrical resistance, capacitance, and tangent loss of the prepared iron ore cuboids were measured and recorded from 300 K to 1100 K using an impedance analyzer (LCR bridge; 0.05% accuracy, 4 Hz–8 MHz, IM3536, Hioki, Nagano, Japan) [30]. A constant voltage mode was employed during this measurement, and temperatures were recorded between 5 kHz–8 MHz. The recorded data (resistance, capacitance, and tangent loss) were transferred to a personal computer via an RS-232C serial port that was connected to the impedance analyzer [30]. The electrical conductivity (σ, $\Omega^{-1}m^{-1}$), dielectric constant (ε'), and dielectric loss (ε'') were calculated using Equations (1), (2), and (3), respectively.

$$\sigma = \frac{t_s}{A * R} \tag{1}$$

$$\varepsilon' = \frac{Cd}{\varepsilon A} \tag{2}$$

$$\varepsilon'' = \varepsilon' \ tan\delta \tag{3}$$

where R (Ω) is the experimentally measured electrical resistance of the studied sample, A is the cross-sectional area of the sample (m^2), t_s is the thickness of the sample (m), C is the experimentally measured electrical capacitance of the sample, $tan\delta$ represents the measured tangent loss, and ε is the free space permittivity.

2.3. Bond Ball Mill Grindability Test

In the mineral industry, the BBMG test is usually employed to estimate the comminution energy, scale up the comminution operation, and compare the material response to ball milling [15]. The test was proposed by Bond in 1961 [31], and ever since, it has been regarded as the industrial standard for estimating the energy requirement for ore milling using the material's index known as the Bond work index (W_i). The grindability test of the studied sample was performed using a laboratory standard Bond ball mill (395-51, BICO Braun International, Burbank, CA, USA) [32]. The sample preparation and BBMG test procedure is presented in Figure 1. A 5 kg representative sample was obtained using a coning and quartering method, and the particle size analysis was performed, as described in Section 2.2. The interstitial volume equals 700 cm^3 within 285 steel balls of different sizes, as described by BICO Braun International, was determined to establish the required feed for the BBMG test [32]. The obtained feed mass (1710.5 g), equivalent to 700 cm^3, was used to determine the particle pack density of the investigated sample. The 285 steel balls (20,125 g) and 1710.5 g samples were fed into the Bond mill, and the cover was lined with rubber material before closing the feed opening to avoid losing particles during milling. The mill was run at 70 rpm for 100 revolutions, after which the product was discharged into a pan. The steel balls were properly cleaned using a brush and re-fed into the mill for further milling operation. The product was batch dry-screened using a laboratory electric sieve shaker (AS 200, Retsch, Retsch-Allee Haan, Germany) by putting the product on a sieve size 106 μm (test sieve size—closing sieve aperture). The undersized product was used to calculate the next number of revolutions that can produce a 250% recirculating load [15]. The representative sample, equivalent to the mass of the undersized product, was added to the recirculating load for the next milling operation. The number of grams per revolution (G_{bp}) was calculated and recorded. This procedure was repeated until G_{bp} becomes constant or changes trend direction. The Bond Equation (4) was employed to calculate the Bond work index of the studied sample.

$$W_i = \frac{44.5}{{P_1}^{0.23} G^{0.82} \left(\frac{10}{\sqrt{P_{80}}} - \frac{10}{\sqrt{F_{80}}} \right)} \tag{4}$$

where W_i represents Bond work index (kWh/t), P_1 is the closing sieve aperture (mm), G represents the ideal grindability (g/rev), P_{80} is the 80% passing size (μm) of product, and F_{80} is the 80% passing size (μm) of feed.

Comparative Method of Grindability Test

Berry and Bruce proposed a comparative method to determine the work index of test samples due to the large sample and time required for the standard Bond grindability test approach [33]. The method was based on the condition that the reference and test samples consume the same energy (E) when ground under the same grinding conditions (sample mass, grinding time, mill, and grinding media). Based on Bond's energy Equation (5), a comparative work index Equation (6) was proposed [33]. When testing ore samples at the same operating conditions, the equation provides a ratio between reference and test ore grinding characteristics. Different researchers have used this approach to calculate the work index of an ore using another ore of known work index [34–36].

$$E = 10W_i\left(\frac{1}{\sqrt{Y_{80}}} - \frac{1}{\sqrt{X_{80}}}\right) \tag{5}$$

$$W_{it} = W_{ir}\frac{\frac{10}{\sqrt{P_r}} - \frac{10}{\sqrt{F_r}}}{\frac{10}{\sqrt{P_t}} - \frac{10}{\sqrt{F_t}}} \tag{6}$$

where W_{ir}—reference work index (obtained using the standard method), Wi_t—test work index, F_r—80% passing size (μm) of reference feed, F_t—80% passing size (μm) of test feed, P_r—80% passing size (μm) of the reference product, and P_t—80% passing size (μm) of test product.

Since the W_i of the as-received sample has been estimated, a comparative method was used to establish the Bond work index (as a measure of grindability) of the microwave-pretreated iron ore samples. To do so, a bulk-treated sample (1710.5 g) was fed into the laboratory standard Bond Ball mill using the same set of steel balls (20,125 g) as that used to perform the grindability test of as-received samples. The sample was ground for 425 runs (based on the average number of runs for the BBMG test of the as-received sample). The product particle size analysis was then performed to obtain P_t. Also, grindability tests were conducted for the separate batch-microwave-treated samples (125 g each to make a total of 1750 g, of which 1710.5 g was sampled for the grindability tests) at 30, 60, and 90 s residence times. The sub-sample of 125 g can be obtained by dividing a 500 g sample into four pieces (see Section 2.5). Since 1710.5 g is required for the BBMG test, 14 sub-samples of 125 g were merged (1750 g). The studied sample cannot be microwaved for longer than 90 s due to the observed increased work index due to fusion of iron ore particles (see Sections 2.5 and 3.4).

2.4. Thermal Pretreatment via Furnace

Thermal pretreatment of the studied samples (Figure 3a) were performed using an industrial furnace with a maximum heating temperature of 1800 °C (Nabertherm VHT 8/22-GR, Lilienthal, Germany; Figure 3b). The studied sample was batch-treated by putting 1710.5 g of the sample in a clay pot placed at the center of the furnace's heating chamber. Heating was performed from room temperature to 150 °C at a heating rate of 10 °C/min. After reaching the target temperature, the sample was heat-shocked for 1 h before cooling to room temperature within the heating chamber. This procedure was repeated at target temperatures of 200, 300, and 400 °C using separate representative samples.

2.5. Thermal Pretreatment via Microwave

Microwave pretreatment of minerals/ores is one of the innovative methods researchers have focused on to improve grinding operation and reduce comminution energy. In this work, a multimode microwave oven (Figure 3c) with cavity dimensions of 381 × 330 × 216 mm (Amana RC17S2, 2.45 GHz frequency, Benton Harbor, MI, USA) was employed for the microwave pretreatment of the prepared representative iron ore samples. The sample to be microwave-treated was prepared by step-wisely dividing a 5 kg sub-sample using a riffle-splitter, which produced approximately 507.5 g. The obtained sub-sample was then reduced to 500 g, as discussed in Section 2.1. The effect of irradiation time on the final temperature reached by the sample was studied using 500 g specimens, each from 30 s up to 180 s at 30 s intervals. At 90 s irradiation time and above, localized burning of iron ore particles was observed, leading to the fusion of particles. For each test, the final bulk temperature was immediately measured after reaching the target heating time using an infrared thermometer with a temperature range of 0–550 °C and reading accuracy of ±3 °C

(Habotest HT650B, Liheng Village, QingXi Town, Dongguan, China). The experiment was repeated three times using separate representative samples (obtained using the same approach as discussed above), and the average final bulk temperature was calculated and recorded. This procedure was repeated using 125 g samples (obtained by dividing 500 g sample into 4 sub-samples using a riffle-splitter) but with a maximum microwave residence time of 90 s, in response to having noticed particles fusing from this temperature upward. The effect of sample quantity on microwave treatment of the studied sample was studied using 125, 250, 500, 1000 (obtained by adding two 500 g sub-samples), and 1500 g (obtained by adding three 500 g sub-samples) specimens at 30 s radiation treatment time. In each case, the average final bulk temperature of three repeated tests was calculated and recorded. To study the effect of microwave pretreatment on the grindability of the selected samples, a representative sample (1710.5 g, sample required for ball milling; see Section 2.1) was bulk treated at 30 s microwave residence time. Also, separate representative samples were batch-treated at 30 s (125 g each to make a total of 1750 g, of which 1710.5 g was sampled for the grindability test). The grindability of the bulk-microwave-treated sample was compared with that of the batch-microwave-treated one. Based on the results obtained, the batch-microwave treatment approach was repeated at 60 and 90 s microwave residence time.

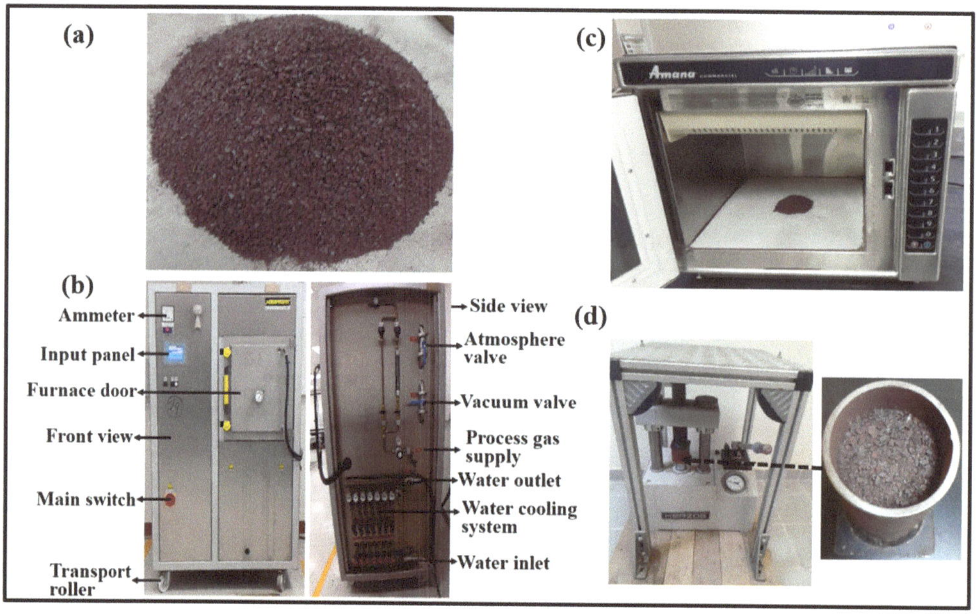

Figure 3. (**a**) Prepared iron ore sample (−3.35 mm), (**b**) industrial furnace (Nabertherm VHT 8/22-GR, Lilienthal, Germany), (**c**) microwave treatment, (**d**) mechanical pretreatment using piston die machine.

2.6. Mechanical Pretreatment

Mechanical pretreatment, as used in this study, is the application of load on crushed rock samples placed inside a pressure pot using a piston-die machine to create cracks in the samples (Figure 3d), which may improve the sample grindability. The compression device used in this study has a force capacity of up to 200 kN (Herzog, TP 20P, 16600 Sprague Road, Suite 400, Cleveland, OH 44130, USA). A pressure pot (cylindrical shape iron container: diameter—53.4 mm, height—90.0 mm) and a support die (diameter—53.3 mm, thickness—9.3 mm) were fabricated and used in this study. The sample mass (441.3 g) filled up to 80.7 mm of the pressure pot was calculated based on its packing density (2.4436 g/cm^3)

and the volume of the container (180.6 cm^3). The effect of applied pressure on the sample displacement was investigated to establish the maximum pressure at which the sample's displacement becomes constant. The support die was placed on the sample, occupying 9.3 mm of the upper space of the pressure pot. The load cell's base was lower and centered on the support die such that the applied load could be uniformly distributed on the sample by slowly jacking the hydraulic pump through its handle (at approximately 10 kN/min). Different confined bed compression tests were performed using applied forces of 10, 20, 30, and 40 kN. In each case, the sample displacement was measured and recorded. The procedure was repeated using separate prepared samples of quartzite. Also, the effect of applied force (at different delay times) on the sample's grindability was studied using separate quartzite samples (due to a shortage of iron ore samples). The packing density of the studied quartzite is 1.8321 g/cm^3 [15] and the volume of the container is 180.6 cm^3, hence, the sub-sample mass for the compaction test is 330.88 g. The stage compaction test (using 330.88 g sample each) was performed at 10 kN until approximately 5 kg of sample was reached—the sample required for the standard BBMG test. This sample was then properly mixed for homogeneity purposes. A BBMG test was performed as described earlier. The procedure was repeated for 20 and 30 kN using separately prepared sub-samples. The effect of load delay time was also investigated at 10, 20, and 30 min using various applied forces (10, 20, and 30 kN). The Bond Ball mill grindability test was performed in each case. Based on piston-die test optimization performed for the quartzite sample, 30 kN, zero-time delay condition was employed for the studied iron ore sample. The sub-sample (441.3 g) was batch-treated (four times) to obtain a total mass of 1765.2 g (of which 1710.5 g was sampled for the grindability test as discussed earlier; see Section 2.1 and Figure 4).

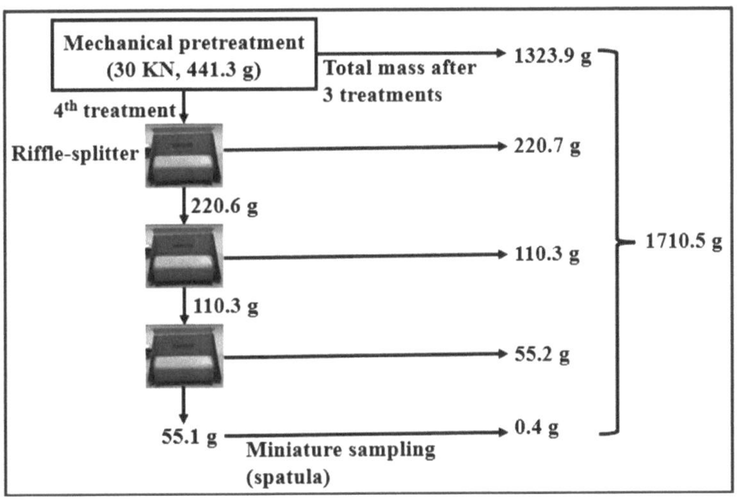

Figure 4. Sub-sampling procedure for hybrid thermal-mechanical-treated iron ore sub-samples.

2.7. Hybrid-Thermal Mechanical Pretreatment

To study the effect of hybrid thermal-mechanical pretreatment on the grindability of the studied samples, only thermal conditions (furnace or microwave) that have the most improved grindability on the studied samples were investigated. Previous studies showed that particles become smaller after a piston die test, which will reduce the impact of microwave radiation on materials' grindability since larger particles respond better to microwave treatment than smaller ones [5]. Therefore, we performed a piston die test of microwave-treated samples and investigated the effect of this process on the grindability of the studied samples. To achieve this, 125 g sub-samples were prepared (Section 2.5) and batch microwave pretreatment was performed (f—2.45 GHz, microwave power—1.7 kW, radiation treatment time—60 s). The process was repeated 15 times to obtain the required sample (1710.5 g) for the BBMG test (Section 2.5). To obtain the sample mass for a piston die test (444.3 g, 30 kN, zero-time delay), the steps in Figure 5 were followed. A riffle-splitter was used to subdivide a 444.3 g thermal-mechanical pretreated sample into appropriate sub-samples and mixed with three times 444.3 g sub-samples. The 1710.5 g sample for the BBBMG test was obtained and the grindability test was performed (Section Comparative Method of Grindability Test). After these processes, the Bond work index of the sample was estimated using the comparative method. Also, the furnace-pretreated sample (heating rate:10 °C; maximum temperature: 400 °C) was mechanically treated (using a piston die device at 30 kN, zero-time delay), and the grindability test was performed using the same grinding condition as the microwave-treated one.

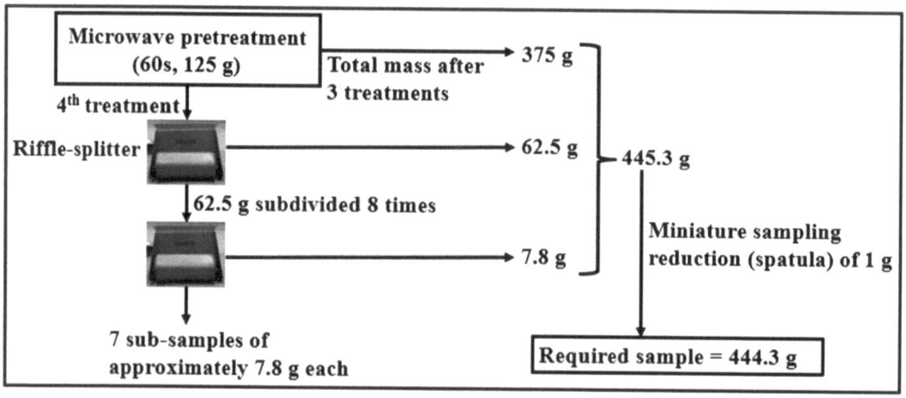

Figure 5. Sub-sampling steps for thermal-mechanical pretreatment of the studied iron ore.

3. Results and Discussions

3.1. FTIR and XRD Analyses

Figure 6 presents the FTIR spectra of the studied iron ore sample. The weak transmittance peak at 563 cm^{-1} is associated with F-O vibration related to hematite [37,38] and magnetite [38]. The FTIR transmittance peak at 785 cm^{-1} is associated with Si-O stretching of silicate minerals [37], while the peak around 873 cm^{-1} is associated with the $CrO4^{-2}$ vibrations—both peaks may be related to diopside mineral [38]. Meanwhile, the peak at around 1430 cm^{-1} is within the carbonate mineral bands, suggesting the vibration of calcium with oxygen [39], which may be linked to diopside. These results suggest that the studied sample contains hematite, magnetite, and diopside. It can be noted that the transmittance peaks of the studied samples remain nearly the same after microwave treatment (Figure 6).

The results of the XRD analysis of the studied sample are presented in Figure 7. Findings indicate that the studied sample contains three minerals: magnetite (Fe_3O_4, card number: 01-076-7161, 3.7% ± 2% by weight), hematite syn (Fe_2O_3, card number: 01-076-4579, 37.3% ± 4% by weight), and diopside (($Mg_{0.845}$ $Fe_{0.048}$ $Al_{0.083}$ $Ti_{0.002}$ $Cr_{0.023}$) ($Ca_{0.828}$ $Na_{0.102}$ $Mg_{0.045}$ $Fe_{0.024}$ $Mn_{0.001}$) (($Si_{0.981}$ $Al_{0.019})_2$ O_6), card number: 01-075-9998, 59% ± 7% by weight). This result is in good agreement with the FTIR analysis. It can be inferred that the selected sample is an iron ore with diopside as the gangue mineral. Diopside may be found with interstitial magnetite and may have different colors, including white, grey, light blue, purple, light to dark green, and maybe multicolor [27]. Other mineral constituents may change the color of the diopside, as in the case of this studied sample (Figure 3a). The effect of microwave irradiation on this type of ore has not been presented in the literature. The crystallographic directions of the mineral phases remain the same after microwave treatment, indicating that microwave irradiation has no significant change in mineral phases of the studied sample. This result agrees with that of the FTIR. The crystallite size and lattice strain of the mineral phases in the investigated untreated sample are presented in Figure 7. Since the crystallite sizes of hematite and diopside have close values, such minerals may be difficult to liberate, resulting in high grinding energy requirements, production of fine particles, and challenges in downstream operations. Previous studies suggested that interlocking of magnetite with diopside may cause metallurgical challenges [27]. In addition, the presence of a small amount of magnetite with fine grain disseminated in the sample's matrix may require more energy to liberate it. Nevertheless, the use of microwave pretreatment for the studied samples improved the grinding process, leading to a reduction in the work index (Section 3.4). This improvement may be attributed to changes in crystallite sizes of the mineral phases and crack propagation on the particles' matrix after microwave pretreatment (see Section 3.2). The XRD analysis showed that the crystallite size of the studied iron ore significantly changed after microwave treatment—the crystallite sizes of hematite and magnetite decreased from 430(6) Å to 26(4) Å and 265(3) Å to 33.82(6) Å, respectively. Meanwhile, that of diopside increased from 438(9) Å to 501.7(9) Å. Further studies are still necessary to determine the effect of microwave irradiation on the mineral liberation of this type of sample and its effect on downstream operations, including the effect on magnetic susceptibility and iron recovery.

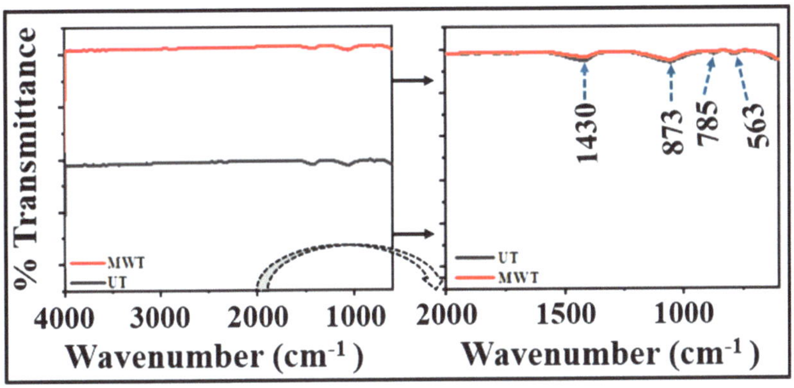

Figure 6. Results of the untreated (UT) and microwave-treated iron ore samples.

3.2. SEM Analysis

The studied sample exhibits a coarse, rough surface texture with various crystallite sizes (Figure 8a–c), which agrees with that of XRD analysis (Figure 7). Minor intergranular and intragranular cracks can be observed on the larger particles after 60 s microwave treatment residence time (Figure 8d,f). After further microwave treatment for up to 90 s, more cracks developed on the particles, as shown in the SEM micrographs (Figure 9a–c). Meanwhile, a fusion of smaller iron particles can be noticed after microwave treatment (Figure 9d–f). This fusion of smaller particles produced bigger ones that take more energy to grind because of strength reinforcement. A similar fusion of particles during microwave treatment had been reported for porphyry copper ore, which increased the Bond work index by 7% [40]. The XRD analysis results indicate changes in minerals' crystallite sizes after microwave treatment. Also, the FTIR transmittance spectra of the studied samples do not change after microwave treatment. However, it may be inferred that the observed fused particles are related to changes in the crystallite sizes of the investigated samples after microwave treatment (see Section 3.1). The results obtained in this study suggest that developed cracks on the larger particles (Figure 9a–c) led to an improved Bond work index of the microwave-treated sample by 6.9%. However, due to the presence of fused particles (Figure 9d–f), the work index at 90 s microwave treatment is higher than that at 60 s microwave treatment (Section 3.4).

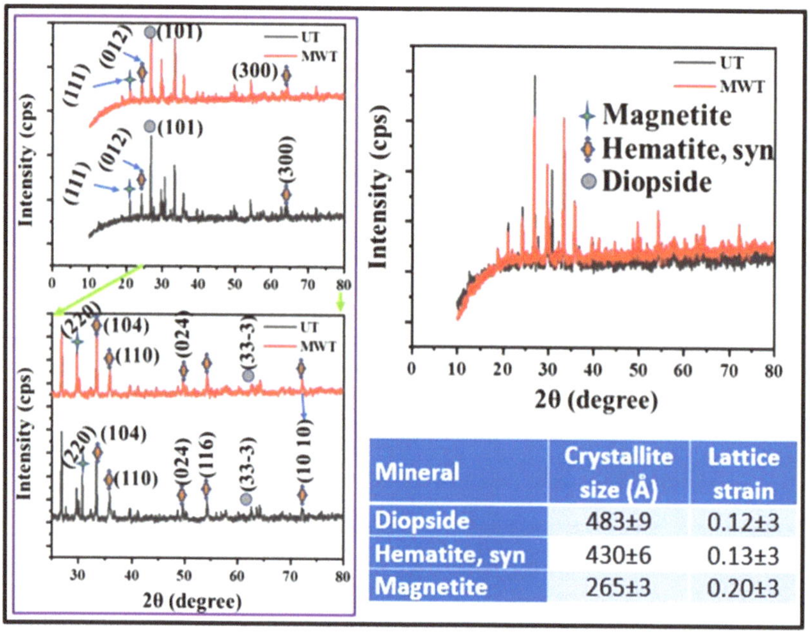

Mineral	Crystallite size (Å)	Lattice strain
Diopside	483±9	0.12±3
Hematite, syn	430±6	0.13±3
Magnetite	265±3	0.20±3

Figure 7. XRD analysis before and after microwave treatment at 1 min microwave irradiation time (UT—untreated sample, MTD—microwave-treated sample for 1 min).

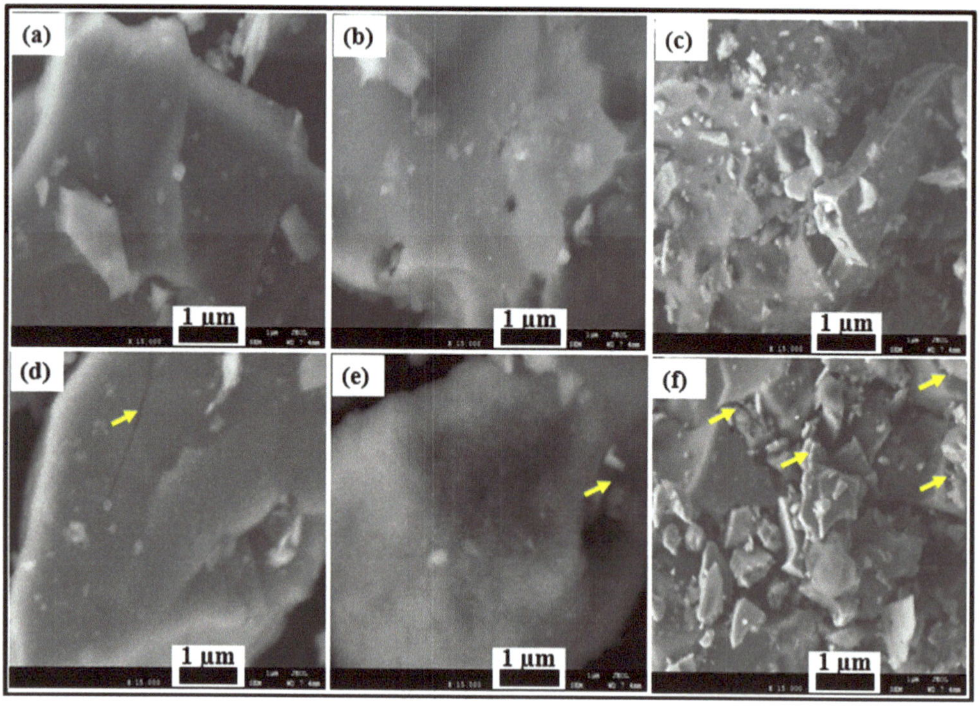

Figure 8. SEM images (15,000×, 5 kV) of the untreated and microwave-treated (2.45 GHz, 1.7 kW, 60 s) samples: (**a**–**c**): untreated, (**d**–**f**): microwave-treated.

Figure 9. (**a–d**) SEM micrographs of the studied microwave-treated (2.45 GHz, 1.7 kW, 90 s) sample; (**e**,**f**) images of fused iron ore particles.

3.3. Electrical Conductivity and Dielectric Properties

The electrical and dielectric properties of iron ore, consisting of magnetite, hematite, and diopside, were analyzed over a temperature range from 300 K to 1100 K. This study examines how temperature affects electrical conductivity and dielectric properties. Figure 10a shows the variation of electrical conductivity (σ) with inverse temperature (1000/T) for different frequencies. The σ increases with temperature for all frequencies, indicating typical semiconducting behavior. At lower frequencies (5 kHz and 10 kHz), a steep decline in σ is observed, suggesting higher thermal activation energy for conduction. As frequency increases, the decrease in σ becomes less steep, indicating reduced thermal activation of charge carriers. At 8 MHz, σ remains relatively stable, reflecting the limited mobility of charge carriers due to the inability to follow the rapidly alternating electric field. Figure 10b shows the dielectric constant (ε') as a function of temperature for various frequencies. The ε' increases significantly with temperature for all frequencies. At 5 kHz and 10 kHz, ε' exhibits a steep increase above 600 K, indicative of enhanced polarization mechanisms like dipolar alignment and interfacial polarization. Higher frequencies (1 MHz and above) show a more gradual increase in ε', suggesting less effective dipolar alignment with the rapidly alternating electric field. The substantial increase in ε' at 5 kHz, reaching around 6×10^5 at approximately 1100 K, underscores strong polarization and dielectric relaxation processes at lower frequencies and elevated temperatures. Figure 10c shows that dielectric loss (ε'') increases with temperature for all frequencies. At lower frequencies (5 kHz and 10 kHz), ε'' rises significantly above 600 K, indicating effective thermal activation of charge carriers and higher energy dissipation. At higher frequencies (above 100 kHz), the increase in ε'' is more gradual. The peak ε'' at 5 kHz reaches about 6×10^7 around 1100 K, highlighting strong ε'' due to conductive and polarization effects from magnetite and hematite. Increased temperature enhances charge carrier mobility, leading to higher ε'' through the alignment and reorientation of dipolar entities in the rock matrix. Figure 10d shows distinct peaks in the loss tangent (tan δ) at various temperatures for different frequencies. At 5 kHz, a prominent peak appears around 850 K, with tan δ values close to 10^5. As frequency increases, these peaks shift to lower temperatures (e.g., 830 K at 10 kHz, 800 K at 20 kHz), indicating a unique relaxation mechanism. This shift suggests that the specific properties of the minerals influence the dielectric relaxation in these rocks. The pronounced peak at 5 kHz around 850 K likely results from dipolar alignment or interfacial polarization. As frequency increases, the relaxation process requires lower thermal energy, causing the temperature shift. The temperature impacts iron ore's electrical and dielectric properties, particularly after 550 K, by enhancing charge carrier activation and dipolar realignment, especially in magnetite and hematite. However, at temperatures lower than 550 K, it seems the electrical and dielectric properties are independent of temperature.

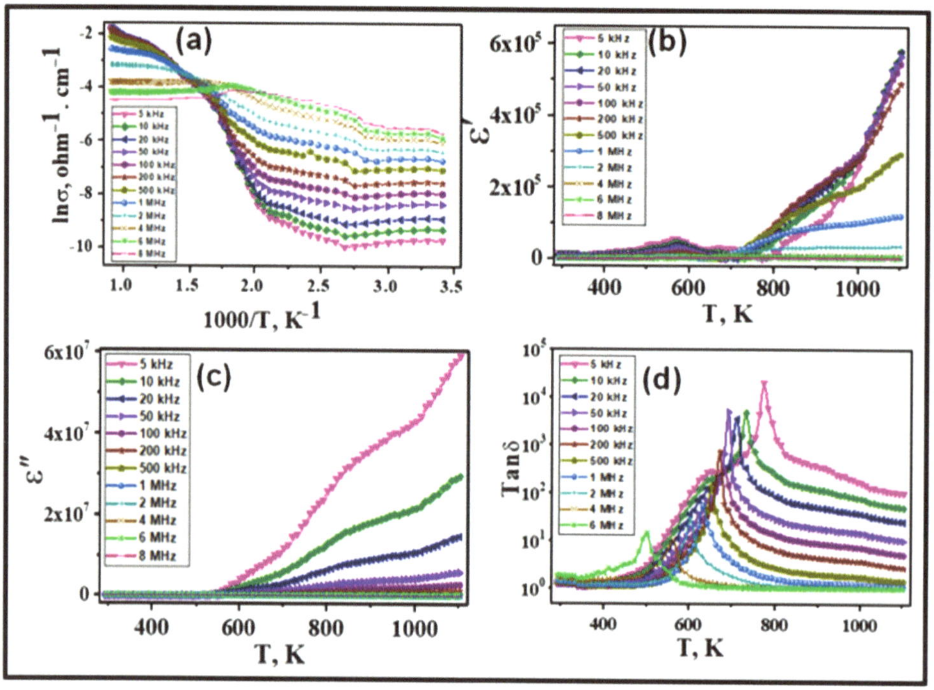

Figure 10. Electrical and dielectric properties of (**a**) electrical conductivity (σ) as a function 1000/T, where T is the absolute sample temperature, (**b**) dielectric constant (e′) (**c**), dielectric loss (e″), and (**d**) dielectric loss tangent (tanδ).

3.4. Effect of Pretreatment Methods on Work Index

The Bond work index of the investigated sample (sieve test size = 106 μm, average pack density = 2.4436 g/cm^3, P_{80} = 91.60 μm, F_{80} = 2742; G(g/rev) = 1.0234 m) is 17.5 KWh/ton. Microwave irradiation of material usually increases the material's temperature. Using 500 g representative samples (see Section 2.5), results indicated that average temperatures increased as the microwave residence time (1.7 kW, 2.45 GHz) increased (Figure 11b). Findings also indicated that the higher the sample mass, the lower the average final temperature reached after microwave treatment (Figure 11a). Based on the findings from Figure 11a,b, a lower sample mass of 125 g was considered for investigating the effect of microwave treatment on the selected iron ore sample (see Section 2.5). Results indicate that the average final temperatures of the investigated samples (125 g, 1.7 kW, 2.45 GHz) at 30, 60, and 90 s microwave irradiation times are 157.5, 258.7, and 302.3 °C, respectively (Figure 11c). As can be noted in Figure 12, the particle size distribution of the studied samples changes at different microwave residence times after grinding using the same milling condition. Figure 12 shows that the P_{80} improved after microwave treatment at 30, 60, and 90 s, leading to improved work indexes by 9.1%, 19.8%, and 6.9%, respectively (Figure 12). Firing and fusion of iron ore particles were noted during microwave treatment at 90 s and beyond. This may be attributed to localized heating and intense, sudden temperature increase of iron ore particles. At 90 s microwave treatment time, the work index of the microwave-treated sample is higher than at 60 s microwave treatment due to greater amount of energy required to liberate fused particles, suggesting that further microwave treatment of the studied sample may continue to increase the work index of the studied sample. Therefore, microwave treatment of the investigated sample beyond 60 s is not desirable. As per the effect of conventional thermal pretreatment on the grindability of the investigated sample, the W_i of the studied sample slightly decreases with increasing temperature, with approximately 4.6% improvement in W_i after 400 °C (Figure 11d). This indicates that conventional thermal pretreatment has no significant effect on the grindability of the studied sample under the investigated temperature range. These results agree with the electric and dielectric properties of the investigated sample since the electrical and dielectric properties are independent of temperature at lower temperatures (Section 3.3).

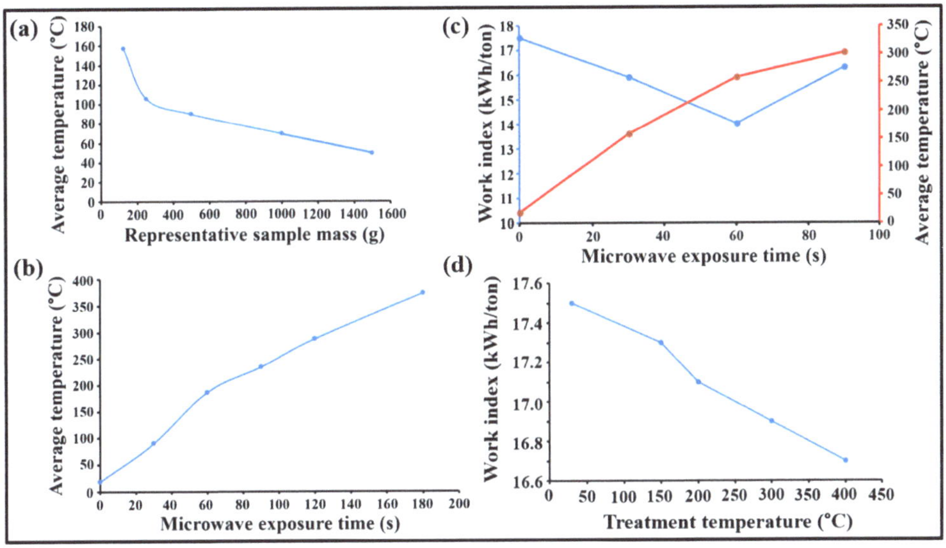

Figure 11. (**a**) Effect of sample mass on microwave treatment of the studied samples, (**b**) Effect of microwave treatment times on iron ore sample's (500 g) response to microwave treatment, (**c**) Work index of the studied iron ore sample (microwave treatment conditions; 2.45 GHz, 1.7 kW, 125 g batch treatment) and average temperatures after microwave treatment, (**d**) Effect of furnace pretreatment on work index of the investigated samples.

To better understand the appropriate sample mass for the mechanical pretreatment test, quartzite and iron ore representative samples were used for sample displacement tests after subjecting them to different applied pressures using a piston die machine. Results show that the higher the applied pressure, the higher the sample displacement (Figure 13a). At 14.86 and 19.81 MPa, the sample displacements nearly have very close values; therefore, 14.86 MPa was regarded as the optimum applied pressure. Due to limited iron ore samples, quartzite samples were used for the optimization of mechanical pretreatment (Figure 13b). Results indicate that the higher the applied pressure, the better the improvement in the W_i of the investigated samples (Figure 13b). Also, the higher the delay time, the better the improvement in the grinding behavior of the studied samples (Figure 13c). The reduction in W_i at 14.86 MPa with zero-time delay is almost the same as that achieved at 4.95 MPa with 30 min delay time (Figure 13b). For this reason, higher applied pressure without delay time was considered for the studied iron ore samples, as delaying ore grinding may affect the downstream process. Findings show that the W_i of the studied iron ore sample after mechanical pretreatment at 14.86 MPa (zero-time delay) is 14.81 kWh/ton, equivalent to approximately 15.4% improvement in the Bond work index. This result suggests that the studied iron ore sample showed a good response to mechanical pretreatment. The combined microwave-mechanical pretreatment (1 min microwave irradiation time, 14.86 MPa, zero-time delay) decreased the W_i of the investigated sample from 17.50 kWh/ton to 13.54 kWh/ton equivalent to 22.6% improvement in its grindability. For the combined conventional thermal-mechanical pretreatment, findings indicate that at 400 °C and 14.86 MPa (zero-time delay), the grindability of the studied sample decreased from 17.50 to 14.14 kWh/ton, equivalent to 19.2% improvement in its Bond work index.

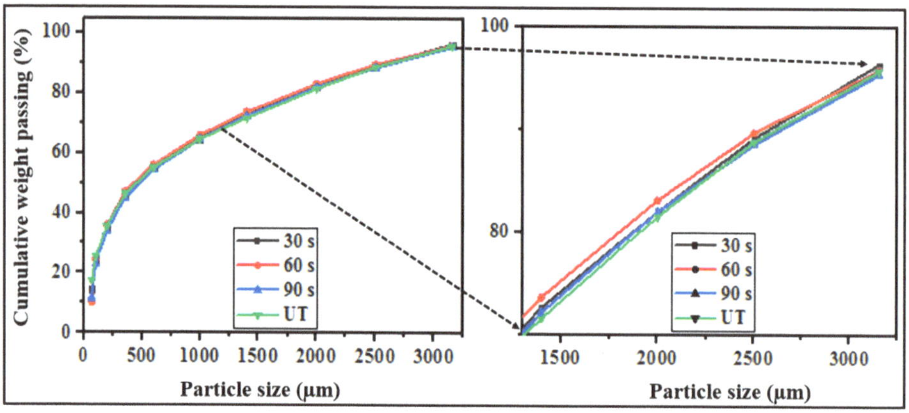

Figure 12. Particle size distribution of untreated (UT—0 s microwave) and microwave-treated iron ore samples after grinding operations.

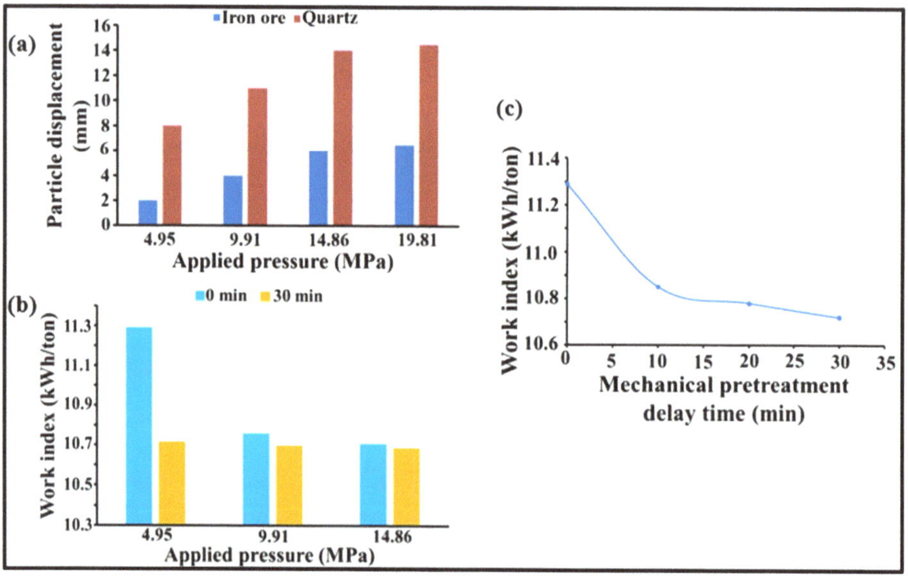

Figure 13. Optimization of mechanical pretreatment test using the quartzite representative samples (reference W_i = 11.49 kWh/ton); (**a**) particles displacement test; (**b**) mechanical pretreatment tests at 0 min and 30 min delay time, using quartzite representative samples; (**c**) effect of delay time on mechanical pretreatment.

4. Conclusions

This paper investigates hybrid thermal and mechanical pretreatments to improve the grinding behavior of fine-grain iron ore (hematite and magnetite interlock with diopside gangue). The mineralogy analysis of the representative sample using the XRD method indicated that the studied sample contain 37.3, 3.7, and 59.0% by weight of hematite, magnetite, and diopside, respectively. The electrical property of the investigated samples showed temperature independence at lower temperatures up to around 550 K, after which the electrical conductivity of the sample increased as the temperature increased. This result agrees with the findings of the grindability tests of pretreated samples using conventional heating since no significant improvement in grindability was achieved below 550 K. At 673 K, only a 4.6% improvement in grindability was achieved. The thermal pretreatment via microwave improved the grindability of the investigated sample up to 19.8% at 60 s microwave irradiation time. Mechanical pretreatment of the sample using a piston die machine at 14.86 MPa (zero-time delay) enhanced the sample's grindability by 15.4%. The hybrid thermal pretreatment was investigated in two ways: conventional-mechanical (400 °C and 14.86 MPa at zero-time delay) and microwave-mechanical (60 s and 14.86 MPa at zero-time delay). Findings indicated that the former improved the grindability of the sample by 19.2% while the latter enhanced the grindability of the sample by 22.6%. These results suggest that stand-alone mechanical pretreatment or microwave pretreatment may be more beneficial in improving the grinding behavior of the studied fine-grain iron ore sample. The results of the mechanical pretreatment obtained in this study may be used in a simulation of the HPGR system for grinding operations of similar iron ore. Future study should focus on economic feasibility of the proposed pretreatment techniques for industrial applications. Among the factors to be considered are input energy of microwave and energy required for mechanical pretreatment.

Author Contributions: Conceptualization, S.O.A.; methodology, S.O.A., H.M.A., H.A.M.A. and A.S.; formal analysis, S.O.A. and A.S.; investigation, S.O.A.; resources, H.A.M.A., H.M.A., A.A., K.L. and R.A.; data curation, S.O.A. and A.S.; writing—original draft preparation, S.O.A., A.A. and A.S.; writing—review and editing, S.O.A., A.S., H.A.M.A. and A.A.; visualization, H.A.M.A., H.M.A., A.A., K.L. and R.A.; supervision, H.A.M.A., H.M.A., A.A. and K.L.; funding acquisition, S.O.A., H.M.A. and H.A.M.A. All authors have read and agreed to the published version of the manuscript.

Funding: This work was supported by the Deanship of Scientific Research (DSR), King Abdulaziz University, Jeddah, under grant No. DG-026-306-1441 and G: 570-135-1441. The authors, therefore, gratefully acknowledge the DSR's technical and financial support.

Data Availability Statement: The original contributions presented in the study are included in the article, further inquiries can be directed to the corresponding author.

Conflicts of Interest: The authors declare no conflicts of interest.

References

1. Jankovic, A.; Valery, W. Reducing Grinding Energy and Cost–Magnetite Iron Ore Design Case Study. In Proceedings of the Conference XII International Mineral Processing Symposium, Cappadocia, Turkey, 6–8 October; 2010; pp. 71–78.
2. Kruyswijk, J.B.; Vlot, E.; Janisch, J. *Mining Engineering*; SME: Englewood, CO, USA, 2024; pp. 16–23.
3. Chryss, A.; Fourie, A.B.; Mönch, A.; Nairn, D.; Seddon, K.D. Towards an Integrated Approach to Tailings Management. *J. South. Afr. Inst. Min. Metall.* **2012**, *112*, 965–969. [CrossRef]
4. Klimpel, R.R. Some Industrial Experiences in Modifying Fine Grinding Environments for Improved Downstream Product Performance. *Int. J. Miner. Process* **1996**, *44*, 133–142. [CrossRef]
5. Adewuyi, S.O.; Ahmed, H.A.M.; Ahmed, H.M.A. Methods of Ore Pretreatment for Comminution Energy Reduction. *Minerals* **2020**, *10*, 423. [CrossRef]
6. Walkiewicz, J.W.; Clark, A.E.; McGill, S.L. Microwave-Assisted Grinding. *IEEE Trans. Ind. Appl.* **1991**, *27*, 239–243. [CrossRef]
7. Liao, N.; Wu, C.; Xu, J.; Feng, B.; Wu, J.; Gong, Y. Effect of Grinding Media on Grinding-Flotation Behavior of Chalcopyrite and Pyrite. *Front. Mater.* **2020**, *7*, 176. [CrossRef]

8. Can, N.M.; Başaran, Ç. Effects of Different Grinding Media and Milling Conditions on the Flotation Performance of a Copper-Pyrite Ore. *Minerals* **2023**, *13*, 85. [CrossRef]
9. Liu, C.; Chen, Z.; Mao, Y.; Yao, Z.; Zhang, W.; Ye, W.; Duan, Y.; Xie, Q. Analysis and Optimization of Grinding Performance of Vertical Roller Mill Based on Experimental Method. *Minerals* **2022**, *12*, 133. [CrossRef]
10. Pamparana, G.; Klein, B.; Bergerman, M.G. Methodology and Model to Predict HPGR Throughput Based on Piston Press Testing. *Minerals* **2022**, *12*, 1377. [CrossRef]
11. Baawuah, E.; Kelsey, C.; Addai-Mensah, J.; Skinner, W. Comparison of the Performance of Different Comminution Technologies in Terms of Energy Efficiency and Mineral Liberation. *Miner. Eng.* **2020**, *156*, 106454. [CrossRef]
12. Ratan, S.; Gupta, R.C.; Dhar, B.B. Ore Grinding Energy Minimisation by Thermal Treatment. In Proceedings of the ICARISM '99 Conference, Perth, WA, USA, 15–17 September 1999.
13. Omran, M.; Fabritius, T.; Mattila, R. Thermally Assisted Liberation of High Phosphorus Oolitic Iron Ore: A Comparison between Microwave and Conventional Furnaces. *Powder Technol.* **2015**, *269*, 7–14. [CrossRef]
14. Han, H.; Lu, L. Thermal Beneficiation of Refractory Iron Ore. In *Iron Ore: Mineralogy, Processing and Environmental Sustainability*; Elsevier: Amsterdam, The Netherlands, 2021; pp. 421–456. ISBN 9780128202265.
15. Adewuyi, S.O.; Ahmed, H.A.M. Grinding Behaviour of Microwave-Irradiated Mining Waste. *Energies* **2021**, *14*, 3991. [CrossRef]
16. Mosher, J.B.; Tague, C.B. Conduct and Precision of Bond Grindability Testing. *Miner. Eng.* **2001**, *14*, 1187–1197. [CrossRef]
17. Kumar, P.; Sahoo, B.K.; De, S.; Kar, D.D.; Chakraborty, S.; Meikap, B.C. Iron Ore Grindability Improvement by Microwave Pre-Treatment. *J. Ind. Eng. Chem.* **2010**, *16*, 805–812. [CrossRef]
18. Song, S.; Campos-Toro, E.F.; López-Valdivieso, A. Formation of Micro-Fractures on an Oolitic Iron Ore under Microwave Treatment and Its Effect on Selective Fragmentation. *Powder Technol.* **2013**, *243*, 155–160. [CrossRef]
19. Singh, V.; Venugopal, R.; Tripathy, S.K.; Saxena, V.K. Comparative Analysis of the Effect of Microwave Pretreatment on the Milling and Liberation Characteristics of Mineral Matters of Different Morphologies. *Miner. Metall. Process.* **2017**, *34*, 65–75. [CrossRef]
20. Abdur Rasyid, M.; Aslam, A.; Rafiei, A.; Sasmito, A.P.; Hassani, F. Transforming Power Draw Trend of Ore Crushing by Applying Microwave Heating. In Proceedings of the 14th International Conference on Applied Energy (ICAE), Bochum, Germany, 8–11 August 2022.
21. Hao, J.; Li, Q.; Qiao, L. Study on Magnetite Ore Crushing Assisted by Microwave Irradiation. *Minerals* **2021**, *11*, 1127. [CrossRef]
22. Liu, L.X.; Powell, M. New Approach on Confined Particle Bed Breakage as Applied to Multicomponent Ore. *Min. Eng.* **2016**, *85*, 80–91. [CrossRef]
23. Abouzeid, A.Z.M.A.; Seifelnassr, A.A.S.; Zain, G.; Mustafa, Y.S. Breakage Behavior of Quartz Under Compression in a Piston Die. *Min. Met. Explor.* **2019**, *36*, 173–180. [CrossRef]
24. Vizcarra, T.G.; Wightman, E.M.; Johnson, N.W.; Manlapig, E.V. The Effect of Breakage Mechanism on the Mineral Liberation Properties of Sulphide Ores. *Min. Eng.* **2010**, *23*, 374–382. [CrossRef]
25. Celik, I.B.; Oner, M. The Influence of Grinding Mechanism on the Liberation Characteristics of Clinker Minerals. *Cem. Concr. Res.* **2006**, *36*, 422–427. [CrossRef]
26. Hosten, C.; Cimilli, H. The Effects of Feed Size Distribution on Confined-Bed Comminution of Quartz and Calcite in Piston-Die Press. *Int. J. Min. Process.* **2009**, *91*, 81–87. [CrossRef]
27. Zhao, S.; Brzozowski, M.J.; Mueller, T.; Wang, L.; Li, W. Classification and Element Mobility in the Yeshan Iron Deposit, Eastern China: Insight from Lithogeochemistry. *Ore. Geol. Rev.* **2022**, *145*, 104909. [CrossRef]
28. Ahmed, H.M.; Ahmed, H.A.M.; Adewuyi, S.O. Characterization of Microschist Rocks under High Temperature at Najran Area of Saudi Arabia. *Energies* **2021**, *14*, 4612. [CrossRef]
29. Akhtar, K.; Khan, S.A.; Khan, S.B.; Asiri, A.M. Scanning Electron Microscopy: Principle and Applications in Nanomaterials Characterization. In *Handbook of Materials Characterization*; Sharma, S.K., Verma, D.S., Khan, L.U., Kumar, S., Khan, S.B., Eds.; Springer: Cham, Switzerland, 2018; pp. 113–145. ISBN 97833199295452.
30. Saeed, A.; Adewuyi, S.O.; Ahmed, H.A.M.; Alharbi, S.R.; Al Garni, S.E.; Abolaban, F. Electrical and Dielectric Properties of the Natural Calcite and Quartz. *Silicon* **2022**, *14*, 5265–5276. [CrossRef]
31. Bond, F.C. Crushing and Grinding Calculations-Part, I. *Br. Chem. Eng.* **1961**, *6*, 378–385.
32. BICO Inc. *Bond Bico Ball Mill Manual*; BICO Inc.: Burbank, CA, USA, 2009.
33. Berry, T.F.; Bruce, R.W. A Simple Method of Determining the Grindability of Ores. *Can. Min. J.* **1996**, *87*, 63–65.
34. Vorster, W.; Rowson, N.A.; Kingman, S.W. The Effect of Microwave Radiation upon the Processing of Neves Corvo Copper Ore. *Int. J. Miner. Process.* **2001**, *63*, 29–44. [CrossRef]
35. Adewuyi, S.O.; Ahmed, H.A.M.; Ahmed, H.M.A. Microwave Treatment of Rocks: Effect on Specific Gravity, Whiteness, and Grindability. *Obogashchenie Rud* **2020**, *3*, 8–13. [CrossRef]
36. Kumar, A.; Kamath, B.P.; Ramarao, V.V.; Mohanty, D.B. Microwave Energy Aided Mineral Comminution. In Proceedings of the International Seminar on Mineral Processing Technology, Chennai, India, 8–10 March 2006; pp. 398–404.
37. Anand, S.; Vinosel, V.M.; Jenifer, M.A.; Pauline, S. Dielectric Properties, AC Electrical Conductivity and Electrical Modulus Profiles of Hematite (α-Fe_2O_3) Nanoparticles. *Int. Res. J. Eng. Technol.* **2017**, *4*, 358–362.
38. Ivashchenko, O.; Jurga-Stopa, J.; Coy, E.; Peplinska, B.; Pietralik, Z.; Jurga, S. Fourier Transform Infrared and Raman Spectroscopy Studies on Magnetite/Ag/Antibiotic Nanocomposites. *Appl. Surf. Sci.* **2016**, *364*, 400–409. [CrossRef]

39. Chukanov, N.V.; Chervonnyi, A.D. *Infrared Spectroscopy of Minerals and Related Compounds*; Springer: Berlin/Heidelberg, Germany, 2016. [CrossRef]
40. Kaya, E. Comminution Behaviour of Microwave Heated Two Sulphide Copper Ores. *Indian J. Chem. Technol.* **2010**, *17*, 455–461.

MDPI AG
Grosspeteranlage 5
4052 Basel
Switzerland
Tel.: +41 61 683 77 34

Minerals Editorial Office
E-mail: minerals@mdpi.com
www.mdpi.com/journal/minerals

www.ingramcontent.com/pod-product-compliance
Lightning Source LLC
Chambersburg PA
CBHW041348210926
51850CB00013B/3020
* 9 7 8 3 7 2 5 8 4 3 0 7 7 *